MEMOIRS
of the
American Mathematical Society

ISSN 0065-9266 (print)
ISSN 1947-6221 (online)

Number 1525

Asymptotic Analysis for Sacks-Uhlenbeck α-Harmonic Maps from Degenerating Riemann Surfaces

Lei Liu
Miaomiao Zhu

November 2024 • Volume 303 • Number 1525 (fifth of 7 numbers)

AMS AMERICAN MATHEMATICAL SOCIETY

Library of Congress Cataloging-in-Publication Data

Cataloging-in-Publication Data has been applied for by the AMS.
See http://www.loc.gov/publish/cip/.
DOI: https://doi.org/10.1090/memo/1525

Memoirs of the American Mathematical Society

This journal is devoted entirely to research in pure and applied mathematics.

Subscription information. Beginning in 2024, Memoirs will be published monthly through 2026. Memoirs is also accessible from www.ams.org/journals. The 2024 subscription begins with volume 293 and consists of twelve mailings, each containing one or more numbers. Individual subscription prices for 2024 are as follows. For electronic only: US$1013. For paper delivery: US$1,163. Add US$22 for delivery within the United States; US$150 for surface delivery outside the United States. Upon request, subscribers to paper delivery of this journal are also entitled to receive electronic delivery. For information on institutional pricing, please visit `https://www.ams.org/publications/journals/subscriberinfo`. Subscription renewals are subject to late fees. See `www.ams.org/journal-faq` for more journal subscription information. Each number may be ordered separately; please specify number when ordering an individual number.

Back number information. For back issues see `www.ams.org/backvols`.

Subscriptions and orders should be addressed to the American Mathematical Society, P. O. Box 845904, Boston, MA 02284-5904 USA. *All orders must be accompanied by payment.* Other correspondence should be addressed to 201 Charles Street, Providence, RI 02904-2213 USA.

Memoirs of the American Mathematical Society (ISSN 0065-9266 (print); 1947-6221 (online)) is published bimonthly (each volume consisting usually of more than one number) by the American Mathematical Society at 201 Charles Street, Providence, RI 02904-2213 USA. Periodicals postage paid at Providence, RI. Postmaster: Send address changes to Memoirs, American Mathematical Society, 201 Charles Street, Providence, RI 02904-2213 USA.

This publication is indexed in *Mathematical Reviews®*, *Zentralblatt MATH*, *Science Citation Index®*, *Science Citation Index™-Expanded*, *ISI Alerting Services℠*, *SciSearch®*, *Research Alert®*, *CompuMath Citation Index®*, *Current Contents®/Physical, Chemical & Earth Sciences.*
This publication is archived in *Portico* and *CLOCKSS*.
Printed in the United States of America.

∞ The paper used in this book is acid-free and falls within the guidelines established to ensure permanence and durability.
Visit the AMS home page at `https://www.ams.org/`

10 9 8 7 6 5 4 3 2 1 29 28 27 26 25 24

Contents

Abstract

In this paper, we investigate the asymptotic analysis and qualitative behaviour for a general sequence of Sacks-Uhlenbeck α-harmonic maps from degenerating Riemann surfaces. This answers an open problem proposed by J. D. Moore, aiming at developing a partial Morse theory for closed parametrized minimal surfaces with arbitrary codimensions in compact Riemannian manifolds. In technical terms, we shall deal with three types of necks which give rise to new difficulties. We establish a generalized energy identity involving some new quantities, such as Pohozaev type constants which measure the extent to which the Pohozaev type identity fails, and some locating constants which characterize the limit position of the blow-up points in long hyperbolic cylinders. Furthermore, we show that the limit of necks are all geodesics in the target manifold where some length formulas are given. Finally, we exploit some geometric and topological conditions to ensure that the limiting neck geodesics appeared in a blow-up sequence of α-harmonic maps with finite Morse index are all of finite length, which implies that energy identity hold.

Received by the editor January 20, 2022, and, in revised form, June 29, 2022.

Article electronically published on November 19, 2024.

DOI: https://doi.org/10.1090/memo/1525

2020 *Mathematics Subject Classification.* Primary 53C43, 58E20.

Key words and phrases. α-harmonic maps, blow-up, energy identity, neck analysis, degenerating Riemann surfaces.

The first author was supported in part by National Natural Science Foundation of China (No. 12101255). The second author was partially supported by National Natural Science Foundation of China (Grant No. 12171314), Innovation Program of Shanghai Municipal Education Commission (Grant No. 2021-01-07-00-02-E00087), and Shanghai Frontier Science Center of Modern Analysis. The second author is the corresponding author.

The first author is affiliated with the School of Mathematics and Statistics, Key Laboratory of Nonlinear Analysis and Applications (Ministry of Education), Hubei Key Laboratory of Mathematical Sciences, Central China Normal University, Wuhan, 430079, People's Republic of China. Email: leiliu2020@ccnu.edu.cn.

The second author is affiliated with the School of Mathematical Sciences, Shanghai Jiao Tong University, 800 Dongchuan Road, Shanghai, 200240, People's Republic of China. Email: mizhu@sjtu.edu.cn.

CHAPTER 1

Introduction

Let (M, g) be a compact Riemannian manifold and let (N, g_N) be a compact Riemannian manifold of dimension n. For a map $u \in C^2(M, N)$, the energy density of u is defined by

$$e(u) = |\nabla_g u|^2 = \text{Trace}_g\ u^*h,$$

where u^*h is the pull-back of the metric tensor h. The energy of the map u is defined as

$$E(u) = \int_M e(u) dvol_g.$$

A critical point of the energy functional $E(\cdot, \cdot)$ is called a harmonic map from (M, g) into (N, g_N). By Nash's embedding theorem, (N, g_N) can be isometrically embedded into some Euclidean space $\mathbb{R}^N$. Then we derive the Euler-Lagrange equations

$$-\Delta_g u = A(u)(\nabla_g u, \nabla_g u),$$

where $A(\cdot, \cdot)$ is the second fundamental form of $N \subset \mathbb{R}^N$ and Δ_g is the Laplace-Beltrami operator on (M, g) which is defined by

$$\Delta_g := \frac{1}{\sqrt{g}} \frac{\partial}{\partial x^\beta} \left(\sqrt{g} g^{\alpha\beta} \frac{\partial}{\partial x^\alpha} \right).$$

Due to the nonlinearity in the Euler-Lagrange equations, which are quadratic in the gradient of the solutions, the existence problem is subtle. Classical variational methods encounter difficulties, since the energy functional $E(\cdot, \cdot)$ does not satisfy the Palais-Smale condition. To overcome these difficulties and to investigate the existence problem of harmonic maps from closed Riemann surfaces, Sacks-Uhlenbeck introduced the following perturbed energy functional in their seminal work [**23**]

$$E_\alpha(\phi, g) = \int_M \{(1 + |d\phi|^2)^\alpha - 1\} dvol_g, \tag{1.1}$$

where $\alpha > 1$ is a constant. Critical points of the functional $E_\alpha(\cdot, \cdot)$, which are called α-harmonic maps, satisfy the following Euler-Lagrange equations:

$$-\Delta_g \phi = (\alpha - 1) \frac{\nabla_g |\nabla_g \phi|^2 \nabla_g \phi}{1 + |\nabla_g \phi|^2} + A(\phi)(d\phi, d\phi). \tag{1.2}$$

The functional (1.1) can be seen as a kind of perturbation of the energy functional $E(\cdot, \cdot)$ that, in contrast to $E(\cdot, \cdot)$ itself, satisfies the Palais-Smale condition. Therefore classical variational scheme can be applied to derive the existence of nontrivial α-harmonic maps.

Furthermore, in [**23**], it was proved that for a sequence of α-harmonic maps with uniformly bounded α-energy, either there exists a nontrivial harmonic sphere or u_α sub-converges smoothly to a limit harmonic map. Later, the finer qualitative bubbling behaviour and asymptotic analysis for a blow-up sequence of α-harmonic maps was explored in various situations. For the energy identity of a minimax sequence of α-harmonic maps, see Jost [**9**] and Lamm [**10**]. For a sequence of minimizing α-harmonic maps, Chen-Tian [**3**] proved that the necks converge to some limit geodesics of finite length which implies that the energy identity holds true. The energy identity for a sequence of α-harmonic maps into certain special target manifolds were proved by Li-Liu-Wang [**13**] and Li-Zhu [**12**]. For a general blow-up sequence of α-harmonic maps, however, the energy identity is not necessary true, see the counter example constructed in [**15**]. Nevertheless, Li-Wang [**14**] investigated the general situation and established a generalized energy identity and studied the asymptotic behaviors of the necks, the limit of which are geodesics in the target manifold but not necessary of finite lengths.

One important application of harmonic maps and α-harmonic maps is the existence theory of minimal surfaces (with or without boundary) in Riemannian manifolds. It is well known that a conformal harmonic map in dimension two is in fact a minimal surface. This provides a way of seeking minimal surfaces (with arbitrary codimensions) in a given Riemannian manifold by finding conformal harmonic maps. This approach was successfully applied by Schoen-Yau [**25**] and Sacks-Uhlenbeck [**23, 24**] for closed surfaces. For each conformal structure on a given closed surface, they produce an energy minimizing map which is a harmonic map, and then minimize over all conformal structures to produce a branched minimal surface. Later, Sacks-Uhlenbeck's scheme was applied to the study of minimal surfaces with free boundary, the case of minimal disks in $\mathbb{R}^3$ was studied by Struwe [**27**], the case of minimal disks in general compact Riemannian manifolds was systematically developed by Fraser [**6**], and the case of higher genus minimal surfaces with free boundary was investigated by Chen-Fraser-Pang [**1**]. Instead of minimizing the Dirichlet energy in a fixed conformal class of the domain surface as in [**24, 25**], they consider Sacks-Uhlenbeck's approach of working with the perturbed energy [**23**]. The advantage of doing this is that they can get some control of the index of the minimal surface by using the Morse theory. Based on the geometric analysis foundation of the blow-up scheme for α-harmonic maps established by Sacks-Uhlenbeck in [**23**] for the interior case and by Fraser in [**6**] for the free boundary case, the key point of above ideas is to understand the asymptotic behaviour for a blow-up sequence of maps when the conformal structures of the underlying domain surface degenerate. In [**1, 24, 25**], some incompressibility assumptions on the fundamental groups of the target manifold were imposed to prevent the degeneration in the process of minimizing over all conformal structures. Generally, Chen-Tian [**3**] studied the asymptotic behaviour for a sequence of minimizing harmonic maps from degenerating surfaces and proved that the limits of necks are some geodesics on the target manifold with finite lengths. For a general sequence of harmonic maps from degenerating surfaces, a generalized energy identity was established in [**28**] by defining Pohozaev type constant associated to harmonic maps from degenerating collar regions and furthermore, Chen-Li-Wang [**4**] showed that the necks appeared in the degenerating collar regions are some geodesics in the target manifold where a length formula for such a geodesic was given. Moreover, some geometric conditions of the target manifold were exploited in [**13**] to ensure that the limiting neck geodesics are of finite lengths so that the energy identity holds. On the other hand, the Morse theory of the α-energy functional also lead to important applications to Riemannian / Kähler geometry and topology, see beautiful and magnificent works by Meeks-Yau [**18**], Siu-Yau [**26**], Micallef-Moore [**17**], Fraser [**7**] and Fraser-Wolfson [**8**] etc.

In order to develop a partial Morse theory for closed parametrized minimal surfaces (with arbitrary codimensions) in compact Riemannian manifolds, Moore proposed an approach (see [**19, 20**]) to find minimal surfaces which is based on the ideas of [**23–25**] but with some modifications. Roughly speaking, in [**23–25**] (or [**1**] for the case of surface with boundary), the idea is firstly to get harmonic maps by minimizing the Dirichlet energy or working with the Sacks-Uhlenbeck's perturbed energy, and then minimize the Dirichlet energy over all conformal structures of the domain surface. While, in [**20**], Moore used Sacks-Uhlenbeck's perturbed energy functional and introduced the notion of *α-minimal surface*, which is a critical point of the following functional:

$$E_\alpha : W^{1,2\alpha}(M,N) \times \mathcal{T}_q \to \mathbb{R},$$

where $\mathcal{T}_q$ is the Teichmüller space of closed Riemann surfaces with genus q. This functional, still denoted by E_α, satisfies Palais-Smale condition for any fixed $\alpha > 1$ which ensures the existence of a critical point, namely, a α-minimal surface. By studying the convergence of a sequence of α-minimal surfaces as $\alpha \searrow 1$, Moore proved that the limit is a conformal harmonic map, i.e. a branched minimal surface, see Chapter 4 in [**20**] for more details. Similarly to [**1, 24, 25**], the key point of this approach is also to understand the asymptotic behaviour of α-harmonic maps as $\alpha \searrow 1$ when the conformal structures of the underlying Riemann surface are tending to the boundary of the Teichmüller space, namely, when the domain Riemann surfaces degenerate. This compactness issue is proposed by Moore as an open problem in his book [**20**], see Problem (2) on page 265 in [**20**].

In this paper, we shall investigate this problem. Now, let (M_n, h_n, c_n) be a sequence of closed hyperbolic Riemann surface of genus $q > 1$, equipped with hyperbolic metrics h_n and compatible complex structure c_n. We consider the compactness problem for a sequence of α-harmonic maps $\{\phi_{\alpha_n}\}$ from (M_n, h_n, c_n) to (N, g_N) with uniformly bounded α-energy

$$\int_{M_n} (1 + |\nabla_{h_n}\phi_{\alpha_n}|^2)^{\alpha_n} dvol_{h_n} \le \Lambda < \infty.$$

Moreover, without loss of generality, we may assume that (M_n, h_n, c_n) degenerates to some limit hyperbolic Riemann surface (M, h, c) by collapsing p $(1 \le p \le 3q-3)$ pairwise disjoint simple closed geodesics γ_n^j of length l_n^j, $j = 1, \dots, p$. For each j, the geodesics γ_n^j degenerate into a pair of punctures $(\varepsilon^{j,1}, \varepsilon^{j,2})$. We refer to [**28**] and references therein for more detailed descriptions.

To analyze the asymptotic behaviour of the above sequence of α_n-harmonic maps $\{\phi_{\alpha_n}\}$, we shall firstly consider the convergence in the domain region which is away from the punctures $(\varepsilon^{j,1}, \varepsilon^{j,2})$, i.e. the domain $\Omega \subset\subset M \setminus \cup_{j=1}^p \{(\varepsilon^{j,1}, \varepsilon^{j,2})\}$. In this case, h_n will converge smoothly to h over Ω. Then by Sacks-Uhlenbeck [**23**], we know that there exist finitely many blow-up points (or energy concentration points) $\{x_1, \dots, x_I\}$, which are away from the punctures $(\varepsilon^{j,1}, \varepsilon^{j,2})$, $j = 1, \dots, p$, such that, by passing to a subsequence, ϕ_{α_n} converges locally and smoothly in the domain region $M \setminus \cup_{j=1}^p \{(\varepsilon^{j,1}, \varepsilon^{j,2})\} \cup \{x_1, \dots, x_I\}$. Next, we need to study the asymptotic behaviour of the sequence of maps near the blow-up points $\{x_1, \dots, x_I\}$ and the punctures $(\varepsilon^{j,1}, \varepsilon^{j,2})$, $j = 1, \dots, p$.

The asymptotic behaviour near the blow-up point x_i, $i \in \{1, 2, \dots, I\}$ in the non degenerating region can be analysed by applying similar arguments as in the case of a fixed domain studied by Li-Wang [**14**] where they established a generalized energy identity and proved that the limit of the necks are some geodesics in the target manifold. Precisely, for a fixed blow-up point x_i, without loss of generality, we may assume there are k_i bubbles occurring at this point, i.e. there are a sequence of points $\{x^{ij}_{\alpha_n}\}$, $j = 1, \dots, k_i$, and a sequence of positive numbers $\{\lambda^{ij}_{\alpha_n}\}$ with $x^{ij}_{\alpha_n} \to x_i$, $\lambda^{ij}_{\alpha_n} \to 0$ as $n \to \infty$ and one of the following two alternatives holds true:

if $1 \le j_1, j_2 \le k_i$ and $j_1 \ne j_2$,

(A1) for any fixed $R > 0$, $D^M_{R\lambda^{ij_1}_{\alpha_n}}(x^{ij_1}_{\alpha_n}) \cap D^M_{R\lambda^{ij_2}_{\alpha_n}}(x^{ij_2}_{\alpha_n}) = \emptyset$, whenever n is sufficiently large.

(A2) $\frac{\lambda^{ij_1}_{\alpha_n}}{\lambda^{ij_2}_{\alpha_n}} + \frac{\lambda^{ij_2}_{\alpha_n}}{\lambda^{ij_1}_{\alpha_n}} \to \infty$, as $n \to \infty$.

Moreover, the following rescaled maps

$$\xi^{ij}_{\alpha_n} := \phi_{\alpha_n}(x^{ij}_{\alpha_n} + \lambda^{ij}_{\alpha_n} x)$$

converge in $C^k_{loc}(\mathbb{R}^2 \setminus \{p^{ij}_1, \dots, p^{ij}_{s_j}\})$ to a nontrivial finite energy harmonic map $\xi_{ij} : \mathbb{R}^2 \to N$, which can be conformally extended to $\mathbb{S}^2$. Usually, we call the above harmonic map ξ^{ij} (or the corresponding scaling factors $(x^{ij}_{\alpha_n}, \lambda^{ij}_{\alpha_n})$) a bubble. To study the asymptotic behavior of the necks, they [**14**] considered the following quantities:

$$\mu_{ij} = \liminf_{n\to\infty} (\lambda^{ij}_{\alpha_n})^{2-2\alpha_n}, \quad \nu_{ij} = \liminf_{n\to\infty} (\lambda^{ij}_{\alpha_n})^{-\sqrt{\alpha_n - 1}}, \tag{1.3}$$

where $\nu_{ij} \in [0, \infty]$, $\mu_{ij} \in [1, \mu_{max}]$ and $\mu_{max} < \infty$. They [**14**] proved the following generalized energy identity that

$$\lim_{\delta\to 0} \lim_{n\to\infty} E_{\alpha_n}(\phi_{\alpha_n}; D_\delta(x_i)) = \sum_{l=1}^{k_i} \mu^2_{il}\, E(\xi^{il}).$$

Furthermore, they showed that the limit of the necks are some geodesics in the target manifold.

In this paper, we shall focus on the asymptotic behaviour of the maps near the punctures $(\varepsilon^{j,1}, \varepsilon^{j,2})$, $j = 1, \dots, p$. By [**28**], we know there is a collar (Q^j_n, g^j_n) near γ^j_n, where

$$Q^j_n = [-T^j_n, T^j_n] \times S^1$$

equipped with the hyperbolic metric

$$g^j_n = (\rho^j_n)^2 ds^2 = (\rho^j_n)^2 (dt^2 + d\theta^2),$$

with

$$\rho^j_n = \frac{l^j_n}{2\pi \cos(\frac{l^j_n t}{2\pi})}, \quad T^j_n = \frac{2\pi}{l^j_n}\left(\frac{\pi}{2} - \arctan(\sinh(\frac{l^j_n}{2}))\right).$$

Then our question can be reduced to the following

PROBLEM. Given a sequence of α_n-harmonic maps $\{\phi_{\alpha_n}\}$ from (Q^j_n, g^j_n) to (N, g_N) with uniformly bounded α_n-energy

$$\int_{M_n} (1 + |\nabla_{h_n}\phi_{\alpha_n}|^2)^{\alpha_n} dvol_{h_n} \le \Lambda < \infty,$$

what is the asymptotic behaviour of $\{\phi_{\alpha_n}\}$ as $\alpha_n \searrow 1$? Specially, can we get some generalized energy identities? What is the limit of these maps over cylinders?

Similarly to the case of harmonic maps, when we study solutions defined on a long cylinder as in above situations, Pohozaev type identities in general do not hold. Therefore, we need to find some geometric quantities associated to the sequence of solutions and cylindrical domains to characterize the asymptotic behaviour of these solutions. Such a scheme was firstly explored in [**28**] for harmonic maps from long cylinders, where the author introduced the Pohozaev type constant by using the Hopf quadratic differential. Such quantities play an important role in the study of the asymptotic behaviour of harmonic maps from degenerating Riemann surfaces [**4, 28**].

Here, in the present paper, we shall introduce the appropriate notion of Pohozaev type constant associated to an α-harmonic map defined on a cylinder, which measures the extent to which the Pohozaev type identity fails.

We associate the sequence (ϕ_{α_n}, M_n) with a sequence of p-tuples $(\kappa_n^1, \dots, \kappa_n^p)$, where

$$\begin{aligned}
\kappa_n^j &= \kappa_n^j(\phi_{\alpha_n})\\
:&= (1-\frac{1}{2\alpha_n})\int_{\{t\}\times S^1}(1+|\nabla_{g_n^j}\phi_{\alpha_n}|^2)^{\alpha_n-1}|\frac{\partial\phi_{\alpha_n}}{\partial t}|^2 d\theta\\
&\quad -\frac{1}{2\alpha_n}\int_{\{t\}\times S^1}(1+|\nabla_{g_n^j}\phi_{\alpha_n}|^2)^{\alpha_n-1}|\frac{\partial\phi_{\alpha_n}}{\partial \theta}|^2 d\theta\\
&\quad -\frac{1}{2\alpha_n}\int_{\{t\}\times S^1}(1+|\nabla_{g_n^j}\phi_{\alpha_n}|^2)^{\alpha_n-1}(\rho_n^j)^2 d\theta\\
&\quad +\frac{1}{\alpha_n}\int_{[0,t]\times S^1}(1+|\nabla_{g_n^j}\phi_{\alpha_n}|^2)^{\alpha_n-1}(\rho_n^j)^3\sin(\frac{l_n^j t}{2\pi})dtd\theta\\
&\quad -(1-\frac{1}{\alpha_n})\int_{[0,t]\times S^1}(1+|\nabla_{g_n^j}\phi_{\alpha_n}|^2)^{\alpha_n-1}|\nabla\phi_{\alpha_n}|^2\rho_n^j\sin(\frac{l_n^j t}{2\pi})dtd\theta.
\end{aligned}\tag{1.4}$$

By Proposition 2.4 in Chapter 2, we know that the quantity κ_n^j is independent of $t\in[-T_n^j, T_n^j]$.

Definition 1.1. We call these constants $\kappa_n^1(\phi_{\alpha_n}), \dots, \kappa_n^p(\phi_{\alpha_n})$ the *Pohozaev type constants* corresponding to the α_n-harmonic map ϕ_{α_n} from Riemann surface M_n.

Remark 1.2. When $\alpha_n \equiv 1$, that is for the harmonic map case, by a direct computation we get that

$$\kappa_n^j = \frac{1}{2}\int_{\{t\}\times S^1}|\frac{\partial\phi_{\alpha_n}}{\partial t}|^2 d\theta - \frac{1}{2}\int_{\{t\}\times S^1}|\frac{\partial\phi_{\alpha_n}}{\partial \theta}|^2 d\theta - \frac{(l_n^j)^2}{4\pi}.$$

One may find that the constant

$$\kappa_n^j + \frac{(l_n^j)^2}{4\pi} = \frac{1}{2}\int_{\{t\}\times S^1}|\frac{\partial\phi_{\alpha_n}}{\partial t}|^2 d\theta - \frac{1}{2}\int_{\{t\}\times S^1}|\frac{\partial\phi_{\alpha_n}}{\partial \theta}|^2 d\theta$$

is just the Pohozaev type constant for harmonic maps which was defined in [**28**] by using the Hopf differential.

Pulling back the hyperbolic metrics h_n and the compatible complex structure c_n by suitable diffeomorphism $M \to M_n \setminus \cup_{j=1}^p \gamma_n^j$ and passing to a subsequence, we may assume (h_n, c_n) all living on the limit surface M and converging in C^∞_{loc}

to (h, c). Thus, ϕ_{α_n} can be seen as a sequence of α_n-harmonic maps defined on (M, h_n, c_n).

To present our main results, we shall first give a general description of the blow-up procedure and the bubbling phenomena. We refer to [**14, 23, 28**] for details on some well known results.

By similar arguments as in the case of harmonic maps [**28**], we can decompose the long cylinder Q_n^j into the concentration parts and the non-concentration parts. However, differently from harmonic maps, since α-harmonic maps are not conformally invariant, here, we need to explore some more detailed information on this kind of decomposition than that in [**28**], especially on the concentration parts.

Roughly speaking, passing to a subsequence, there exists a constant $K_j > 0$ independent of n and $2K_j$ sequences $\{a_n^{j,1}\}$, $\{b_n^{j,1}\}$, $\{a_n^{j,2}\}$, $\{b_n^{j,2}\}$,..., $\{a_n^{j,K_j}\}$, $\{b_n^{j,K_j}\}$, such that

$$-T_n^j \leq a_n^{j,1} \ll b_n^{j,1} \leq a_n^{j,2} \ll b_n^{j,2} \leq \cdots \leq a_n^{j,K_j} \ll b_n^{j,K_j} \leq T_n^j$$
$$(a_n^{j,i} \ll b_n^{j,i} \text{ means } \lim_{n\to\infty} b_n^{j,i} - a_n^{j,i} = \infty)$$

and $-T_n^j \ll a_n^{j,1}$, $b_n^{j,K_j} \ll T_n^j$, $(b_n^{j,i} - a_n^{j,i}) \ll T_n^j$, i.e.

(1.5)

$$\lim_{n\to\infty} a_n^{j,1} - (-T_n^j) = \infty, \quad \lim_{n\to\infty} T_n^j - b_n^{j,K_j} = \infty \ , \lim_{n\to\infty} \frac{b_n^{j,i} - a_n^{j,i}}{T_n^j} = 0, i = 1, \ldots, K_j.$$

Denote

$$J_n^{j,k} := [a_n^{j,k}, b_n^{j,k}] \times S^1, \ k = 1, \ldots, K_j,$$
$$I_n^{j,0} := [-T_n^j, a_n^{j,1}] \times S^1, \ I_n^{j,K_j} := [b_n^{j,K_j}] \times S^1,$$
$$I_n^{j,i} := [b_n^{j,i}, a_n^{j,i+1}] \times S^1, \ i = 1, \ldots, K_j - 1.$$

Then

(1) $\forall i = 0, 1, \ldots, K_j$,

$$\lim_{n\to\infty} \sup_{t\in I_n^{j,i}} \int_{[t,t+1]\times S^1} |\nabla \phi_{\alpha_n}|^2 = 0. \tag{1.6}$$

The maps ϕ_{α_n} on $I_n^{j,i}$ are necks corresponding to collapsing homotopically nontrivial curves.

(2) $\forall k = 1, \ldots, K_j$, we can find finite blow-up points (or energy concentration points) in $J_n^{j,k}$ denoted by $x_n^{j,k;1}, \ldots, x_n^{j,k;m_{jk}}$, where m_{jk} is a positive integer independent of n, such that

$$\left| x_n^{j,k;m} - (\frac{a_n^{j,k} + b_n^{j,k}}{2}, 0) \right| \leq C, \ m = 1, \ldots, m_{jk}, \tag{1.7}$$

$$\limsup_{n\to\infty} E(\phi_{\alpha_n}, D_r(x_n^{j,k;m})) \geq \epsilon_0', \quad \forall \, r > 0, \ m = 1, \ldots, m_{jk}$$

where $\epsilon_0' > 0$ is the constant defined in Lemma 3.4 which is depending only on N, and there exists a harmonic map $\psi_{jk} : (-\infty, +\infty) \times S^1 \to N$ such that

$$\begin{gathered} \overline{\phi}_{\alpha_n}(x) := \phi_{\alpha_n}\left((\frac{a_n^{j,k} + b_n^{j,k}}{2}, 0) + x \right) \to \psi_{jk} \\ \text{strongly in } C^1_{loc}\left([-R, R] \times S^1 \setminus S_{jk} \right), \ \forall R > 0 \end{gathered} \tag{1.8}$$

and

$$\lim_{R\to\infty}\lim_{n\to\infty} E\left(\overline{\phi}_{\alpha_n}, (-\frac{b_n^{j,k}-a_n^{j,k}}{2}, -R)\times S^1\right) + \lim_{R\to\infty}\lim_{n\to\infty} E\left(\overline{\phi}_{\alpha_n}, (R, \frac{b_n^{j,k}-a_n^{j,k}}{2})\times S^1\right) = 0, \tag{1.9}$$

where S_{jk} is the limit of the following translated blow-up point set (by passing to a subsequence)

$$S_{jk}^n := \left\{x_n^{j,k;1}, \ldots, x_n^{j,k;m_{jk}}\right\} - \left(\frac{b_n^{j,k}+a_n^{j,k}}{2}, 0\right).$$

Note that this limit exists due to the property (1.7).

For the sake of convenience, we shall define this kind of harmonic map

$$\psi_{jk} : (-\infty, +\infty)\times S^1 \to N$$

appearing in the blow-up process near the punctures as follows:

DEFINITION 1.3. We call ψ_{jk} a weak limit of the sequence ϕ_{α_n} near the jth collar γ_n^j (or near the punctures $(\varepsilon^{j,1}, \varepsilon^{j,2})$), $j = 1, \ldots, p$.

Here we would like to remark that ψ_{jk} might be a trivial map, i.e., a constant map (and thus it is sometimes called a ghost bubble). If not, then by classical theory of harmonic maps, ψ_{jk} can be conformally extended to a non-trivial harmonic sphere.

Near each blow-up point of S_{jk}, we will obtain some bubbles, i.e., harmonic maps from S^2 to N. This process is standard now and we refer to [**14, 23**] for details. For simplicity of notations, we assume S_{jk} is a single point set and $S_{jk} = \{0\}$. By [**14, 23**], we can construct a bubble tree structure. See also the beginning of Chapter 3. Precisely, we assume there are L_{jk} bubbles, i.e. there are a sequence of points $\{x_{\alpha_n}^{j,k;l}\}$, $l = 1, \ldots, L_{jk}$, and a sequence of positive numbers $\{\lambda_{\alpha_n}^{j,k;l}\}$ with $x_{\alpha_n}^{j,k;l} \to 0$, $\lambda_{\alpha_n}^{j,k;l} \to 0$ as $n\to\infty$ and one of two alternatives $(A1)$ and $(A2)$ holds. Moreover, the rescaled maps

$$w_{\alpha_n}^{j,k;l} := \phi_{\alpha_n}(x_{\alpha_n}^{j,k;l} + \lambda_{\alpha_n}^{j,k;l}x)$$

will converge weakly in $W_{loc}^{1,2}(\mathbb{R}^2)$ to a nontrivial finite energy harmonic map $w_{jk}^l : \mathbb{R}^2 \to N$.

To study the asymptotic behavior of α-harmonic maps from degenerating Riemann surfaces, besides defining the quantities considered as in [**14**]

$$\mu_{jk}^l = \liminf_{n\to\infty} (\lambda_{\alpha_n}^{j,k;l})^{2-2\alpha_n}, \quad \nu_{jk}^l = \liminf_{n\to\infty} (\lambda_{\alpha_n}^{j,k;l})^{-\sqrt{\alpha_n - 1}}, \tag{1.10}$$

we need to define the following two new quantities:

$$\liminf_{n\to\infty} \sqrt{\Xi_n^j}\frac{\pi^3}{l_n^j}, \quad l_{jk}^\infty = \liminf_{n\to\infty}\left(\frac{l_n^j}{2\pi\cos(\frac{l_n^j}{2\pi}\frac{a_n^{j,k}+b_n^{j,k}}{2})}\right)^{2-2\alpha_n},$$

where

$$\Xi_n^j := \frac{1}{2\alpha_n - 1}\left(4\alpha_n\kappa_n^j(\phi_{\alpha_n}) + \frac{(l_n^j)^2}{\pi}\right).$$

Here, we remark that the first quantity Ξ_n^j corresponds to the Pohozaev type constant κ_n^j given in Definition 1.1 and the second quantity l_{jk}^∞ is a kind of locating constant which characterizes the blow-up position in the cylindrical collar (Q_n^j, g_n^j).

Now, we will state our first main theorem about a generalized energy identity for α-harmonic maps from degenerating Riemann surfaces.

THEOREM 1.4. *Assumptions and notations as above. Suppose*

$$\liminf_{n\to\infty}\max_{j=1,\dots,p}(l_n^j)^{2(1-\alpha_n)}\le\Lambda_1<\infty.$$

Then there exist at most finitely many blow-up points $\{x_1,\dots,x_I\}$ which are away from the punctures $\{(\epsilon^{j,1},\epsilon^{j,2}),j=1,2,\dots,p\}$ and finitely many harmonic maps

- $\phi:(\overline{M},\overline{c})\to N$, *where $(\overline{M},\overline{c})$ is the normalization of (M,c),*
- $\xi_{il}:S^2\to N$, $l=1,\dots,k_i$, *near the blow-up point x_i, $i=1,\dots,I$,*
- $\psi_{jk}:S^2\to N$, $k=1,\dots,K_j$, *are some weak limits near the punctures $\{(\epsilon^{j,1},\epsilon^{j,2})\}$, $j=1,2,\dots,p$,*
- $w_{jk}^l:S^2\to N$, $l=1,\dots,L_{jk}$, *near the weak limit ψ_{jk} and also near the punctures $\{(\epsilon^{j,1},\epsilon^{j,2})\}$, $k=1,\dots,K_j$, $j=1,2,\dots,p$,*

such that after selection of a subsequence, $\phi_{\alpha_n}\to\phi$ in the sense of $C^\infty_{loc}(M\setminus\{x_1,\dots,x_I\})$ and the following generalized energy identity hold:

$$\begin{aligned}\lim_{n\to\infty}E_{\alpha_n}(\phi_{\alpha_n},h_n)=&E(\phi)+\sum_{i=1}^{I}\sum_{l=1}^{k_i}\mu_{il}^2E(\xi_{il})+\sum_{j=1}^{p}\sum_{k=1}^{K_j}l_{jk}^\infty E(\psi_{jk})\\&+\sum_{j=1}^{p}\sum_{k=1}^{K_j}l_{jk}^\infty\sum_{l=1}^{L_{jk}}(\mu_{jk}^l)^2E(w_{jk}^l)+\sum_{j=1}^{p}\lim_{n\to\infty}\kappa_n^j\frac{2\pi^2}{l_n^j}.\end{aligned}$$

Here, ψ_{jk} may be a trivial map, i.e. a constant map (which is also called a ghost bubble), while ξ_{il} and w_{jk}^l are nontrivial maps.

We would like to remark that the constants l_{jk}^∞, μ_{il}, μ_{jk}^l given in Theorem 1.4 are bounded from above and satisfying

$$1\le l_{jk}^\infty\le\max\left\{\Lambda_1,\frac{\Lambda}{c}\right\},\ 1\le\mu_{il}\le\frac{\Lambda}{c},\ 1\le\mu_{jk}^l\le\frac{\Lambda}{c},$$

where $c>0$ is defined as the minimal energy of nontrivial harmonic spheres into N, i.e.

$$c:=\inf\left\{\ E(u)\mid u:S^2\to N \text{ is a nontrivial harmonic map }\right\}.$$

In fact, since

$$\pi l_n^j\le 2\pi\cos\left(\frac{l_n^j}{2\pi}t\right)\le 2\pi,$$

it is easy to see that

$$1\le l_{jk}^\infty\le(l_n^j)^{2-2\alpha_n}\le\Lambda_1.$$

For the following bounds

$$1\le l_{jk}^\infty\le\frac{\Lambda}{c},\ 1\le\mu_{il}\le\frac{\Lambda}{c},\ 1\le\mu_{jk}^l\le\frac{\Lambda}{c},$$

see the beginning of Chapter 3.

Our next main theorem is to study the asymptotic properties of the necks between these bubbles and weak limits.

THEOREM 1.5. *Under the assumptions and notations of Theorem 1.4, suppose that*

$$\liminf_{n\to\infty} \frac{(1-\alpha_n)\log l_n}{(l_n)^2} \le \Lambda < \infty,$$

then the images between $\phi(M)$, the weak limits $\psi_{jk}(S^2)$ and the bubbles $\xi_{il}(S^2)$, $w^l_{jk}(S^2)$ are connected by some geodesics on N.

Moreover, we have:

(1) *If there is only one bubble ξ_{i1} (i.e. $k_i = 1$) near the blow-up point x_i, $i = 1, \ldots, I$, then $\xi_{i1}(S^2)$ connects to $\phi(M)$ by a geodesic on N with the length*

$$\sqrt{\frac{E(\xi_{i1})}{\pi}} \log \nu_{i1}.$$

(2) *If there is only one bubble w^1_{jk} (i.e. $L_{jk} = 1$) near the weak limit ψ_{jk}, $k = 1, \ldots, K_j$, $j = 1, \ldots, p$, then $w^1_{jk}(S^2)$ connects to $\psi_{jk}(S^2)$ by a geodesic on N with the length*

$$\sqrt{\frac{E(w^1_{jk})}{\pi}} \log \nu^1_{jk}.$$

(3) *For each $j = 1, \ldots, p$, the asymptotic limit of the sum of the lengths of the geodesics connecting $\phi(M)$ and $\psi_{jk}(S^2)$, $k = 1, \ldots, K_j$ near the jth pair of punctures is*

$$\lim_{n\to\infty} \sqrt{\Xi^j_n} \frac{\sqrt{\pi^3}}{l^j_n}, \quad j = 1, \ldots, p.$$

According to Theorem 1.4, it is easy to see that the energy identity holds if and only if the following analytical conditions are satisfied

$$\mu_{il} = 1, \quad l^{\infty}_{jk} = 1, \quad \mu^l_{jk} = 1, \quad \lim_{n\to\infty} \kappa^j_n \cdot \frac{2\pi^2}{l^j_n} = 0.$$

A natural question then is whether we can exploit some geometric and topological conditions to ensure that the limiting necks are some geodesics of finite length so that the energy identity follows immediately.

To achieve this, we shall impose the assumptions that the Ricci curvature of the target manifold has a positive lower bound or the target manifold has a finite fundamental group, and consider a sequence of α-harmonic maps with bounded Morse index (see Section 2.5). Analogous assumptions were imposed in e.g. [**13**,**20**].

In the following, we state our third main result:

THEOREM 1.6. *Under the assumption of Theorem 1.5, suppose that the sequence $\{\phi_{\alpha_n}\}$ has bounded index and one of the following two assumptions holds:*

(i) *the Ricci curvature of the target manifold (N, g_N) has a positive lower bound, i.e. there exists a positive constant $\lambda_N > 0$ such that $Ric_N \ge \lambda_N > 0$;*
(ii) *the target manifold (N, g_N) has a finite fundamental group.*

Then the limit of all necks consists of geodesics of finite length. Moreover, the following energy identity holds, i.e.

(1.11)

$$\lim_{n\to\infty} E_{\alpha_n}(\phi_{\alpha_n}, h_n) = E(\phi) + \sum_{i=1}^{I}\sum_{l=1}^{k_i} E(\xi_{il}) + \sum_{j=1}^{p}\sum_{k=1}^{K_j} E(\psi_{jk}) + \sum_{j=1}^{p}\sum_{k=1}^{K_j}\sum_{l=1}^{L_{jk}} E(w_{jk}^l).$$

Since α-harmonic maps are not conformally invariant, the asymptotic analysis and qualitative behaviour for a sequence of α-harmonic maps from degenerating Riemann surfaces is much more complicated than the case of α-harmonic maps from non-degenerating Riemann surfaces e.g. [**6, 10, 14, 23**] and the cases of harmonic maps from non-degenerating or degenerating Riemann surfaces e.g. [**3, 5, 9, 21, 22, 28**].

To investigate and describe the asymptotic properties in the above limiting process, we shall firstly introduce the following three types of necks in the blow-up analysis:

- The first type neck: we define first two kinds of necks corresponding to the collapsing geodesics γ_n^j or punctures $(\varepsilon^{j,1}, \varepsilon^{j,2})$, $j = 1, \ldots, p$. As discussed before, near each γ_n^j, there is a standard cylinder collar (Q_n^j, g_n^j) which can be decomposed into energy concentration parts $J_n^{j,k}$, $k = 1, \ldots, K_j$ and non-concentration parts $I_n^{j,i}$, $i = 0, 1, \ldots, K_j$. We call the above energy non-concentration parts $I_n^{j,i}$, $i = 0, 1, \ldots, K_j$ the first type neck.
- The second type neck: in the energy concentration parts $J_n^{j,k}$, $k{=}1, \ldots, K_j$, as discussed before, there exist finitely many blow-up points and at each such point, we can get a neck domain in the classical sense (see [**5**] or the definition of the third type neck below). The difference here is that the underlying Riemannian metric over the collar region is degenerating. Since the Euler-Lagrange equations of α-harmonic maps and the α-energy E_α are not conformally invariant and also in order to distinguish this type of neck from the third type neck, we call this one the second type neck.
- The third type neck: we define the third type neck as in the classical sense [**5**], i.e. near the blow-up point x_i, $i = 1, \ldots, I$ which is away from the punctures. Precisely, for each fixed x_i, there are k_i bubbles $(x_{\alpha_n}^{ij}, \lambda_{\alpha_n}^{ij})$, $j = 1, \ldots, k_i$ and for simplicity of notations, we assume $k_i = 1$, i.e. there is only one bubble at the blow-up point x_i. Then the rescaled map $\phi_{\alpha_n}(x_{\alpha_n}^{i1} + \lambda_{\alpha_n}^{i1} x)$ converges strongly in $W_{loc}^{1,2}(\mathbb{R}^2)$ to a harmonic map ξ_{i1} and we call the domain $D_\delta(x_{\alpha_n}^{i1}) \setminus D_{\lambda_{\alpha_n} R}(x_{\alpha_n}^{i1})$ the third type neck in this paper. Note that the domain metric is not degenerating in the third type neck.

Secondly, we will encounter some technical difficulties due to the facts that α-harmonic maps are not conformally invariant and the domain metric is degenerating. For instance, we need to prove small energy regularity results on the first type and second type necks, which are based on the structures of the Euler-Lagrange equations and some computations of the domain metric. How to derive a Pohozaev type identity on the second type neck and how to construct a Pohozaev type constant on the first type neck? Especially for the later question since we do not have the Hopf differential as in the harmonic map case considered in [**5, 28**]. Here, we will use the method of integrating by parts which is based on some observation

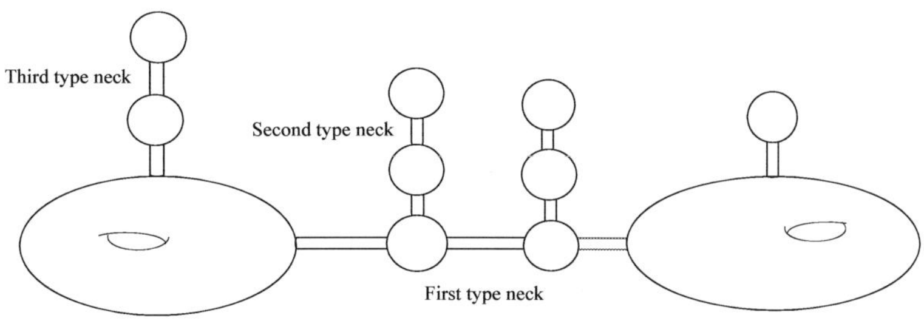

FIGURE 1. Illustation of three types of necks

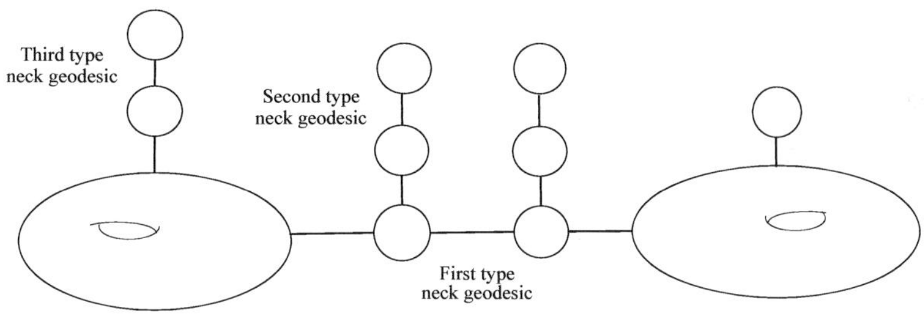

FIGURE 2. Illustation of three types of limiting neck geodesics

of the structure of Euler-Lagrange equations. In formulas of Pohozaev type identity and Pohozaev type constant, there are some new additional terms appearing. How to control these terms on the first and second type neck which is in fact the main difficulty part of this paper. One will find that there are some new quantities appearing in the argument.

The rest of the paper is organized as follows. In Chapters 2 and 3, we will study the limit process on the first and second type neck respectively. We will prove a generalized energy identity and show that the limit of the neck are some geodesics on the target manifold where the length formulas are given. Besides this, we will exploit some geometric conditions to ensure that the limiting necks are geodesics of finite length so that energy identities hold. In Chapter 4, we will prove our main Theorems 1.4–1.6.

CHAPTER 2

Asymptotic analysis of first type neck

In this chapter, we will study the asymptotic behaviour of the first type neck for α-harmonic maps as described in Chapter 1.

For reader's convenience, we conclude the detailed problems at the beginning of this section which will be studied later. Recalling that there is a standard cylinder collar (Q_n^j, g_n^j) near the jth closed geodesic γ_n^j, $j = 1, \dots, p$. For simplicity of notations, we omit the subscript j here. Let $Q_n = [-T_n, T_n] \times S^1$ with the metric $g_n := \rho_n^2 ds^2 = \rho_n^2(dt^2 + d\theta^2)$ where

$$\rho_n = \frac{l_n}{2\pi\cos(\frac{l_n t}{2\pi})}$$

and

$$T_n = \frac{2\pi}{l_n}\left(\frac{\pi}{2} - \arctan(\sinh(\frac{l_n}{2}))\right) \to \infty$$

as $n \to \infty$. Then the asymptotic behaviour of maps on the first type neck is equivalent to studying the limit on the following sub-cylinder

$$(T_n^1, T_n^2) \times S^1 \subset Q_n,$$

where $-T_n + 2 \leq T_n^1 < T_n^2 \leq T_n - 2$ and

$$\lim_{n\to\infty} \sup_{t\in[T_n^1, T_n^2]} E(\phi_{\alpha_n}, [t-1, t+1]) = 0.$$

Roughly speaking, we will study the following question in this section: Given a sequence of α_n-harmonic maps ϕ_{α_n} from Q_n to N with uniformly bounded α-energy

$$\int_{Q_n} (1 + |\nabla_{g_n}\phi_{\alpha_n}|^2)^{\alpha_n} dvol_{g_n} \leq \Lambda < \infty,$$

does ϕ_{α_n} converges to a geodesic on the cylinder $(T_n^1, T_n^2) \times S^1$? If so, how to compute the lengths of these curves?

2.1. Some results on the first type neck

In this subsection, we will state some theorems on the first type neck, which will be proved in this part. We firstly present the following theorem about a generalized energy identity.

THEOREM 2.1. *Let ϕ_{α_n} be a sequence of α_n-harmonic maps from (Q_n, g_n) to N with*

$$\liminf_{n\to\infty} (l_n)^{2(1-\alpha_n)} \leq C < \infty$$

and with uniformly bounded α-energy

$$\int_{Q_n} (1 + |\nabla_{g_n}\phi_{\alpha_n}|^2)^{\alpha_n} dvol_{g_n} \leq \Lambda < \infty,$$

which satisfies

$$\lim_{n\to\infty}\sup_{t\in[T_n^1,T_n^2]} E(\phi_{\alpha_n},[t-1,t+1])=0,$$

where $-T_n+2\le T_n^1<T_n^2\le T_n-2$, *then there holds*

$$\lim_{n\to\infty}\int_{[T_n^1,T_n^2]\times S^1}\left((1+|\nabla_{g_n}\phi_{\alpha_n}|^2)^{\alpha_n}-1\right)dvol_{g_n}=\lim_{n\to\infty}2\kappa_n(\phi_{\alpha_n})|T_n^2-T_n^1|.$$

Our second main theorem is studying the limit of the first type neck, which is a geodesic of the target manifold. Precisely, we will prove

THEOREM 2.2. *Let* ϕ_{α_n} *be a sequence of* α_n*-harmonic maps from* Q_n *to* N *with uniformly bounded* α*-energy*

$$\int_{Q_n}(1+|\nabla_{g_n}\phi_{\alpha_n}|^2)^{\alpha_n}dvol_{g_n}\le\Lambda<\infty,$$

which satisfies

$$\lim_{n\to\infty}\sup_{t\in[T_n^1,T_n^2]} E(\phi_{\alpha_n},[t-1,t+1])=0$$

and

$$\lim_{n\to\infty}\frac{(1-\alpha_n)\log l_n}{(l_n)^2}<\infty,$$

where $-T_n+2\le T_n^1<T_n^2\le T_n-2$. *Denote*

$$\Xi_n:=\frac{1}{2\alpha_n-1}\left(4\alpha_n\kappa_n(\phi_{\alpha_n})+\frac{l_n^2}{\pi}\right)$$

and

$$\Xi:=\lim_{n\to\infty}\sqrt{|\Xi_n|}(T_n^2-T_n^1),$$

then we have

(1) *when* $\Xi=0$, $\{\phi_{\alpha_n}\}$ *converges to a point;*
(2) *when* $\Xi\in(0,\infty)$, $\{\phi_{\alpha_n}\}$ *converges to a geodesic of length* $\sqrt{\frac{1}{4\pi}}\Xi$;
(3) *when* $\Xi=\infty$, $\{\phi_{\alpha_n}\}$ *converges to a geodesic of infinite length.*

Our third main theorem on the first type neck is to exploit some geometric conditions to ensure that the limiting neck are geodesics of finite length so that the energy identity holds.

THEOREM 2.3. *Let* (N,g_N) *be a compact Riemannian manifold with one of the following two assumptions holds:*

(i) *the Ricci curvature of the target manifold* (N,g_N) *has a positive lower bound, i.e. there exists a positive constant* $\lambda_N>0$ *such that* $Ric_N\ge\lambda_N>0$;
(ii) *the target manifold* (N,g_N) *has a finite fundamental group.*

Let ϕ_{α_n} *be a sequence of* α_n*-harmonic maps from* (Q_n,g_n) *to* N *with uniformly bounded* α*-energy and Morse index, i.e.*

$$\int_{Q_n}(1+|\nabla_{g_n}\phi_{\alpha_n}|^2)^{\alpha_n}dvol_{g_n}\le\Lambda<\infty,\quad Index\ (\phi_{\alpha_n};E_{\alpha_n})\le\Lambda_{Index}<\infty,$$

which satisfies

$$\lim_{n\to\infty}\sup_{t\in[T_n^1,T_n^2]} E(\phi_{\alpha_n},[t-1,t+1]\times S^1)=0$$

and

$$\lim_{n\to\infty} \frac{(1-\alpha_n)\log l_n}{(l_n)^2} < \infty, \tag{2.1}$$

where $-T_n+2 \le T_n^1 < T_n^2 \le T_n - 2$, *then passing to a subsequence,* $\{\phi_{\alpha_n}\}$ *converges to a geodesic of finite length* $\sqrt{\frac{1}{4\pi}\Xi} < \infty$.

Moreover, the following energy identity holds,

$$\lim_{n\to\infty} \int_{[T_n^1,T_n^2]\times S^1} \left((1+|\nabla_{g_n}\phi_{\alpha_n}|^2)^{\alpha_n} - 1\right) dvol_{g_n} = 0.$$

The rest part of Chapter 2 is organized as follows. In Section 2.2, we will introduce some basic lemmas on the first type neck, such as the Pohozaev type constant, small energy regularity and three circle type method. In Section 2.3, we will establish a generalized energy identity, i.e. Theorem 2.1. In Section 2.4, we will study some refined neck analysis on the first type neck and then prove Theorem 2.2. The proof of Theorem 2.3 will be given in Section 2.5.

2.2. Some basic lemmas

In this subsection, we will give some basic lemmas, such as the Pohozaev type constant, small energy regularity and three circle lemmas.

It is easy to see that the equation (1.2) has the following two equivalent forms:

$$\Delta\phi_\alpha + (\alpha-1)\frac{\nabla|\nabla_g\phi_\alpha|^2\nabla\phi_\alpha}{1+|\nabla_g\phi_\alpha|^2} + A(\phi_\alpha)(d\phi_\alpha, d\phi_\alpha) = 0, \tag{2.2}$$

$$\text{div }\left\{(1+|\nabla_g\phi_\alpha|^2)^{\alpha-1}\nabla\phi_\alpha\right\} + (1+|\nabla_g\phi_\alpha|^2)^{\alpha-1}A(\phi_\alpha)(d\phi_\alpha, d\phi_\alpha) = 0. \tag{2.3}$$

Here and in the sequel, the operators Δ, ∇, div are defined as the standard operators for Euclidean domains (D, dx^2+dy^2), e.g. $\Delta = \frac{\partial^2}{\partial x^2} + \frac{\partial^2}{\partial y^2}$, or for standard cylindrical domains $((T_1,T_2)\times S^1, dt^2+d\theta^2)$, e.g. $\Delta = \frac{\partial^2}{\partial t^2} + \frac{\partial^2}{\partial \theta^2}$.

We start from a lemma about the Pohozaev type constant which is a key quantity in the formula of a generalized energy identity and the length of limiting neck geodesic.

PROPOSITION 2.4. *Let* ϕ_{α_n} *be a* α_n*-harmonic map from* (Q_n, g_n) *to* N. *Then for any* $t \in [-T_n, T_n]$,

$$\begin{aligned}
\kappa_n(\phi_{\alpha_n}) := &(1-\frac{1}{2\alpha_n})\int_{\{t\}\times S^1}(1+|\nabla_{g_n}\phi_{\alpha_n}|^2)^{\alpha_n-1}|\frac{\partial\phi_{\alpha_n}}{\partial t}|^2 d\theta \\
&- \frac{1}{2\alpha_n}\int_{\{t\}\times S^1}(1+|\nabla_{g_n}\phi_{\alpha_n}|^2)^{\alpha_n-1}|\frac{\partial\phi_{\alpha_n}}{\partial \theta}|^2 d\theta \\
&- \frac{1}{2\alpha_n}\int_{\{t\}\times S^1}(1+|\nabla_{g_n}\phi_{\alpha_n}|^2)^{\alpha_n-1}\rho_n^2 d\theta \\
&+ \frac{1}{\alpha_n}\int_{[0,t]\times S^1}(1+|\nabla_{g_n}\phi_{\alpha_n}|^2)^{\alpha_n-1}\rho_n^3\sin(\frac{l_n t}{2\pi})dtd\theta \\
&- (1-\frac{1}{\alpha_n})\int_{[0,t]\times S^1}(1+|\nabla_{g_n}\phi_{\alpha_n}|^2)^{\alpha_n-1}|\nabla\phi_{\alpha_n}|^2\rho_n\sin(\frac{l_n t}{2\pi})dtd\theta
\end{aligned} \tag{2.4}$$

is independent of $t \in [-T_n, T_n]$*, where*

$$\int_{[0,t]\times S^1} f dt d\theta = -\int_{[t,0]\times S^1} f dt d\theta, \ if \ t < 0.$$

PROOF. Denote

$$Q_{t_1 t_2} := [t_1, t_2] \times S^1.$$

Multiplying (2.3) by $\frac{\partial \phi_{\alpha_n}}{\partial t}$ and integrating by parts, we get

$$\begin{aligned}
0 &= \int_{Q_{t_1 t_2}} div \left\{ (1 + |\nabla_{g_n} \phi_{\alpha_n}|^2)^{\alpha_n - 1} \nabla \phi_{\alpha_n} \right\} \frac{\partial \phi_{\alpha_n}}{\partial t} dt d\theta \\
&= \int_{\partial Q_{t_1 t_2}} (1 + |\nabla_{g_n} \phi_{\alpha_n}|^2)^{\alpha_n - 1} |\frac{\partial \phi_{\alpha_n}}{\partial t}|^2 d\theta \\
&\quad - \int_{Q_{t_1 t_2}} (1 + |\nabla_{g_n} \phi_{\alpha_n}|^2)^{\alpha_n - 1} \nabla \phi_{\alpha_n} \nabla \frac{\partial \phi_{\alpha_n}}{\partial t} dt d\theta \\
&= \int_{\partial Q_{t_1 t_2}} (1 + |\nabla_{g_n} \phi_{\alpha_n}|^2)^{\alpha_n - 1} |\frac{\partial \phi_{\alpha_n}}{\partial t}|^2 d\theta \\
&\quad - \frac{1}{2} \int_{Q_{t_1 t_2}} (1 + |\nabla_{g_n} \phi_{\alpha_n}|^2)^{\alpha_n - 1} \frac{\partial}{\partial t} |\nabla \phi_{\alpha_n}|^2 dt d\theta. \qquad (2.5)
\end{aligned}$$

Noting that

$$|\nabla \phi_{\alpha_n}|^2 = \rho_n^2 |\nabla_{g_n} \phi_{\alpha_n}|^2 \quad and \quad \frac{d\rho_n}{dt} = \rho_n^2 \sin(\frac{l_n t}{2\pi})$$

we obtain

$$\begin{aligned}
&- \frac{1}{2} \int_{Q_{t_1 t_2}} (1 + |\nabla_{g_n} \phi_{\alpha_n}|^2)^{\alpha_n - 1} \frac{\partial}{\partial t} |\nabla \phi_{\alpha_n}|^2 dt d\theta \\
&= -\frac{1}{2} \int_{Q_{t_1 t_2}} (1 + |\nabla_{g_n} \phi_{\alpha_n}|^2)^{\alpha_n - 1} \frac{\partial}{\partial t} |\nabla_{g_n} \phi_{\alpha_n}|^2 \rho_n^2 dt d\theta \\
&\quad - \int_{Q_{t_1 t_2}} (1 + |\nabla_{g_n} \phi_{\alpha_n}|^2)^{\alpha_n - 1} |\nabla_{g_n} \phi_{\alpha_n}|^2 \rho_n \frac{d\rho_n}{dt} dt d\theta \\
&= -\frac{1}{2\alpha_n} \int_{Q_{t_1 t_2}} \frac{\partial}{\partial t} (1 + |\nabla_{g_n} \phi_{\alpha_n}|^2)^{\alpha_n} \rho_n^2 dt d\theta \\
&\quad - \int_{Q_{t_1 t_2}} (1 + |\nabla_{g_n} \phi_{\alpha_n}|^2)^{\alpha_n - 1} |\nabla_{g_n} \phi_{\alpha_n}|^2 \rho_n^3 \sin(\frac{l_n t}{2\pi}) dt d\theta \\
&= -\frac{1}{2\alpha_n} \int_{\partial Q_{t_1 t_2}} (1 + |\nabla_{g_n} \phi_{\alpha_n}|^2)^{\alpha_n} \rho_n^2 d\theta \\
&\quad + \frac{1}{\alpha_n} \int_{Q_{t_1 t_2}} (1 + |\nabla_{g_n} \phi_{\alpha_n}|^2)^{\alpha_n} \rho_n^3 \sin(\frac{l_n t}{2\pi}) dt d\theta \\
&\quad - \int_{Q_{t_1 t_2}} (1 + |\nabla_{g_n} \phi_{\alpha_n}|^2)^{\alpha_n - 1} |\nabla_{g_n} \phi_{\alpha_n}|^2 \rho_n^3 \sin(\frac{l_n t}{2\pi}) dt d\theta. \qquad (2.6)
\end{aligned}$$

Combining (2.5) with (2.6), we have

$$
\begin{aligned}
0 =& \int_{\partial Q_{t_1 t_2}} (1+|\nabla_{g_n}\phi_{\alpha_n}|^2)^{\alpha_n-1}|\frac{\partial\phi_{\alpha_n}}{\partial t}|^2 d\theta - \frac{1}{2\alpha_n}\int_{\partial Q_{t_1 t_2}} (1+|\nabla_{g_n}\phi_{\alpha_n}|^2)^{\alpha_n}\rho_n^2 d\theta \\
&+ \frac{1}{\alpha_n}\int_{Q_{t_1 t_2}} (1+|\nabla_{g_n}\phi_{\alpha_n}|^2)^{\alpha_n}\rho_n^3 \sin(\frac{l_n t}{2\pi}) dt d\theta \\
&- \int_{Q_{t_1 t_2}} (1+|\nabla_{g_n}\phi_{\alpha_n}|^2)^{\alpha_n-1}|\nabla_{g_n}\phi_{\alpha_n}|^2 \rho_n^3 \sin(\frac{l_n t}{2\pi}) dt d\theta \\
=& (1-\frac{1}{2\alpha_n})\int_{\partial Q_{t_1 t_2}} (1+|\nabla_{g_n}\phi_{\alpha_n}|^2)^{\alpha_n-1}|\frac{\partial\phi_{\alpha_n}}{\partial t}|^2 d\theta \\
&- \frac{1}{2\alpha_n}\int_{\partial Q_{t_1 t_2}} (1+|\nabla_{g_n}\phi_{\alpha_n}|^2)^{\alpha_n-1}|\frac{\partial\phi_{\alpha_n}}{\partial \theta}|^2 d\theta \\
&- \frac{1}{2\alpha_n}\int_{\partial Q_{t_1 t_2}} (1+|\nabla_{g_n}\phi_{\alpha_n}|^2)^{\alpha_n-1}\rho_n^2 d\theta \\
&+ \frac{1}{\alpha_n}\int_{Q_{t_1 t_2}} (1+|\nabla_{g_n}\phi_{\alpha_n}|^2)^{\alpha_n-1}\rho_n^3 \sin(\frac{l_n t}{2\pi}) dt d\theta \\
&- (1-\frac{1}{\alpha_n})\int_{Q_{t_1 t_2}} (1+|\nabla_{g_n}\phi_{\alpha_n}|^2)^{\alpha_n-1}|\nabla\phi_{\alpha_n}|^2 \rho_n \sin(\frac{l_n t}{2\pi}) dt d\theta,
\end{aligned}
$$

which immediately implies the conclusion of this proposition. □

Secondly, we will prove the following small energy regularity lemma. This type of result was firstly introduced by Sacks-Uhlenbeck [**23**] for a Euclidean domain. Since we shall mainly focus on the hyperbolic cylindrical domain (Q_n, g_n) for neck analysis and α-harmonic map is not conformally invariant, here, we present a detailed proof of this hyperbolic cylindrical version.

LEMMA 2.5. *Let ϕ_{α_n} be an α_n-harmonic map from (Q_n, g_n) to N. Then there exist positive constants $\epsilon_0 > 0$ and $\alpha_0 > 1$ depending only on N, such that, for any $t \in (-T_n + 2, T_n - 2)$, if $1 < \alpha_n < \alpha_0$ and*

$$E(\phi_{\alpha_n}, [t-2, t+2] \times S^1) \leq \epsilon_0,$$

then

$$\|\nabla\phi_{\alpha_n}\|_{C^1([t-1,t+1]\times S^1)} \leq C(N)\|\nabla\phi_{\alpha_n}\|_{L^2([t-2,t+2]\times S^1)}.$$

PROOF. Firstly, we introduce a new coordinate system. Let (r, θ) be polar coordinates centered at 0. Let

$$F : \mathbb{R}^1 \times S^1 \to \mathbb{R}^2, F(t,\theta) = (e^t, \theta), (t,\theta) \in \mathbb{R}^1 \times S^1 \tag{2.7}$$

where $\mathbb{R}^1 \times S^1$ is equipped with the metric $\rho_n^2 ds^2 = \rho_n^2(dt^2 + d\theta^2)$, which is conformal to the standard Euclidean metric $g_{\mathbb{R}^2}$ on $\mathbb{R}^2$. In fact,

$$(F^{-1})^*\rho_n^2 ds^2 = \frac{\rho_n^2}{r^2} g_{\mathbb{R}^2}.$$

We denote

$$\Phi_{\alpha_n} := \phi_{\alpha_n} \circ F^{-1}. \tag{2.8}$$

Then it is easy to see that Φ_{α_n} satisfies

$$\Delta_{h_n}\Phi_{\alpha_n} = -(\alpha_n - 1)\frac{\nabla_{h_n}|\nabla_{h_n}\Phi_{\alpha_n}|^2\nabla_{h_n}\Phi_{\alpha_n}}{1+|\nabla_{h_n}\Phi_{\alpha_n}|^2} + A(\nabla_{h_n}\Phi_{\alpha_n}, \nabla_{h_n}\Phi_{\alpha_n}), \quad \text{in } D_{e^{t+2}}(0)\setminus D_{e^{t-2}}(0), \tag{2.9}$$

where $h_n = \frac{\rho_n^2}{r^2}g_{\mathbb{R}^2}$. This equation is equivalent to the following equation that

$$\begin{aligned}\Delta\Phi_{\alpha_n} &= A(\nabla\Phi_{\alpha_n}, \nabla\Phi_{\alpha_n}) - (\alpha_n - 1)\frac{\nabla|\nabla_{h_n}\Phi_{\alpha_n}|^2\nabla\Phi_{\alpha_n}}{1+|\nabla_{h_n}\Phi_{\alpha_n}|^2}\\
&= A(\nabla\Phi_{\alpha_n}, \nabla\Phi_{\alpha_n}) - (\alpha_n - 1)\frac{\nabla(\frac{r^2}{\rho_n^2}|\nabla\Phi_{\alpha_n}|^2)\nabla\Phi_{\alpha_n}}{1+\frac{r^2}{\rho_n^2}|\nabla\Phi_{\alpha_n}|^2}\\
&= A(\nabla\Phi_{\alpha_n}, \nabla\Phi_{\alpha_n}) - (\alpha_n - 1)\frac{\frac{r^2}{\rho_n^2}\nabla|\nabla\Phi_{\alpha_n}|^2\nabla\Phi_{\alpha_n}}{1+\frac{r^2}{\rho_n^2}|\nabla\Phi_{\alpha_n}|^2}\\
&\quad - (\alpha_n - 1)\frac{\nabla(\frac{r^2}{\rho_n^2})|\nabla\Phi_{\alpha_n}|^2\nabla\Phi_{\alpha_n}}{1+\frac{r^2}{\rho_n^2}|\nabla\Phi_{\alpha_n}|^2}.\end{aligned}$$

Without loss of generality, we may assume $t = 0$ and

$$\int_{D_{e^2}(0)\setminus D_{e^{-2}}(0)}\Phi_{\alpha_n}dx = 0.$$

Take a cut-off function $\eta \in C_0^\infty(D_{e^2}(0)\setminus D_{e^{-2}}(0))$ such that

$$0 \le \eta \le 1, \quad \eta|_{D_e(0)\setminus D_{e^{-1}}(0)} = 1, \quad |\nabla\eta| + |\nabla^2\eta| \le C.$$

Noting that

$$\left|(\alpha_n - 1)\frac{\frac{r^2}{\rho_n^2}\nabla|\nabla\Phi_{\alpha_n}|^2\nabla\Phi_{\alpha_n}}{1+\frac{r^2}{\rho_n^2}|\nabla\Phi_{\alpha_n}|^2}\right| \le C(\alpha_n - 1)|\nabla^2\Phi_{\alpha_n}|$$

and

$$\begin{aligned}&\left|(\alpha_n - 1)\frac{\nabla(\frac{r^2}{\rho_n^2})|\nabla\Phi_{\alpha_n}|^2\nabla\Phi_{\alpha_n}}{1+\frac{r^2}{\rho_n^2}|\nabla\Phi_{\alpha_n}|^2}\right|\\
&= \left|(\alpha_n - 1)\frac{\frac{2r}{\rho_n^2}|\nabla\Phi_{\alpha_n}|^2\frac{\partial\Phi_{\alpha_n}}{\partial r} - \frac{2r}{\rho_n}\sin(\frac{l_n\log r}{2\pi})|\nabla\Phi_{\alpha_n}|^2\frac{\partial\Phi_{\alpha_n}}{\partial r}}{1+\frac{r^2}{\rho_n^2}|\nabla\Phi_{\alpha_n}|^2}\right|\\
&\le C(\alpha_n - 1)\left(\frac{|\frac{\partial\Phi_{\alpha_n}}{\partial r}|}{r} + \frac{|\rho_n\sin(\frac{l_n\log r}{2\pi})||\frac{\partial\Phi_{\alpha_n}}{\partial r}|}{r}\right)\\
&\le C(\alpha_n - 1)\frac{|\frac{\partial\Phi_{\alpha_n}}{\partial r}|}{r}\end{aligned}$$

where we used the fact that

$$\frac{d\rho_n}{dr} = \frac{\rho_n^2}{r}\sin\left(\frac{l_n\log r}{2\pi}\right), \quad |\rho_n(t)| \le C$$

for all $t \in [-T_n, T_n]$ and $\alpha_n > 1$, computing directly, we get

$$\begin{aligned}
|\Delta(\eta\Phi_{\alpha_n})| &= |\eta\Delta\Phi_{\alpha_n} + 2\nabla\eta\nabla\Phi_{\alpha_n} + \Delta\eta\Phi_{\alpha_n}| \\
&\le C(N)\left(|\nabla\Phi_{\alpha_n}||\eta\nabla\Phi_{\alpha_n}| + (\alpha_n - 1)|\eta\nabla^2\Phi_{\alpha_n}| + |\nabla\Phi_{\alpha_n}| + |\Phi_{\alpha_n}|\right) \\
&\le C(N)\left(|\nabla\Phi_{\alpha_n}||\nabla(\eta\Phi_{\alpha_n})| + (\alpha_n - 1)|\nabla^2(\eta\Phi_{\alpha_n})| + |\nabla\Phi_{\alpha_n}| + |\Phi_{\alpha_n}|\right). \qquad (2.10)
\end{aligned}$$

By standard elliptic estimates, Sobolev inequality and Poincaré inequality, for any $1 < p < 2$, we get

$$\begin{aligned}
\|\eta\Phi_{\alpha_n}\|_{W^{2,p}(D_{e^2}(0))} &\le C\left(\|\Delta(\eta\Phi_{\alpha_n})\|_{L^p(D_{e^2}(0))} + \|\eta\Phi_{\alpha_n}\|_{L^p(D_{e^2}(0))}\right) \\
&\le C\| \, |\nabla(\eta\Phi_{\alpha_n})| \cdot |\nabla\Phi_{\alpha_n}| \, \|_{L^p(D_{e^2}(0))} \\
&\quad + C(\alpha_n - 1)\|\nabla^2(\eta\Phi_{\alpha_n})\|_{L^p(D_{e^2}(0)\setminus D_{e^{-2}}(0))} \\
&\quad + C\|\nabla\Phi_{\alpha_n}\|_{L^p(D_{e^2}(0)\setminus D_{e^{-2}}(0))} + C\|\Phi_{\alpha_n}\|_{L^p(D_{e^2}(0)\setminus D_{e^{-2}}(0))} \\
&\le C\|\nabla\Phi_{\alpha_n}\|_{L^2(D_{e^2}(0))}\|\nabla(\eta\Phi_{\alpha_n})\|_{L^{\frac{2p}{2-p}}(D_{e^2}(0))} \\
&\quad + C(\alpha_n - 1)\|\nabla^2(\eta\Phi_{\alpha_n})\|_{L^p(D_{e^2}(0)\setminus D_{e^{-2}}(0))} \\
&\quad + C\|\nabla\Phi_{\alpha_n}\|_{L^p(D_{e^2}(0)\setminus D_{e^{-2}}(0))} + C\|\Phi_{\alpha_n}\|_{L^p(D_{e^2}(0)\setminus D_{e^{-2}}(0))} \\
&\le C\|\nabla\Phi_{\alpha_n}\|_{L^2(D_{e^2}(0))}\|\nabla(\eta\Phi_{\alpha_n})\|_{W^{1,p}(D_{e^2}(0))} \\
&\quad + C(\alpha_n - 1)\|\nabla^2(\eta\Phi_{\alpha_n})\|_{L^p(D_{e^2}(0)\setminus D_{e^{-2}}(0))} \\
&\quad + C\|\nabla\Phi_{\alpha_n}\|_{L^p(D_{e^2}(0)\setminus D_{e^{-2}}(0))}.
\end{aligned}$$

Taking first $p = \frac{4}{3}$, if $\epsilon_0 > 0$ and $\alpha_0 - 1 > 0$ is small enough, we have

$$\|\nabla(\eta\Phi_{\alpha_n})\|_{W^{1,\frac{4}{3}}(D_{e^2}(0))} \le C(N)\|\nabla\Phi_{\alpha_n}\|_{L^2(D_{e^2}(0)\setminus D_{e^{-2}}(0))}$$

which implies

$$\|\nabla\Phi_{\alpha_n}\|_{L^4(D_{e^{3/2}}(0)\setminus D_{e^{-3/2}}(0))} \le C(N)\|\nabla\Phi_{\alpha_n}\|_{L^2(D_{e^2}(0)\setminus D_{e^{-2}}(0))}.$$

Choosing a new cut-off function $\eta_1 \in C_0^\infty(D_{e^{3/2}}(0) \setminus D_{e^{-3/2}}(0))$ such that

$$0 \le \eta_1 \le 1, \quad \eta_1|_{D_{e^{5/4}}(0)\setminus D_{e^{-5/4}}(0)} = 1, \quad |\nabla\eta_1| + |\nabla^2\eta_1| \le C,$$

using (2.10), Poincaré inequality and $W^{2,2}$-estimate for the Laplace operator, we have

$$\begin{aligned}
&\|\nabla(\eta_1\Phi_{\alpha_n})\|_{W^{1,2}(D_{e^2}(0))} \\
&\le C\left(\|\Delta(\eta_1\Phi_{\alpha_n})\|_{L^2(D_{e^2}(0))} + \|\nabla\Phi_{\alpha_n}\|_{L^2(D_{e^2}(0)\setminus D_{e^{-2}}(0))}\right) \\
&\le C\Big((\alpha_n - 1)\|\nabla^2(\eta_1\Phi_{\alpha_n})\|_{L^2(D_{e^2}(0))} + \||\nabla\Phi_{\alpha_n}|^2\|_{L^2(D_{e^{3/2}}(0)\setminus D_{e^{-3/2}}(0))} \\
&\quad + \|\nabla\Phi_{\alpha_n}\|_{L^2(D_{e^2}(0)\setminus D_{e^{-2}}(0))}\Big) \\
&\le C\left((\alpha_n - 1)\|\nabla^2(\eta_1\Phi_{\alpha_n})\|_{L^2(D_{e^2}(0))} + \|\nabla\Phi_{\alpha_n}\|_{L^2(D_{e^2}(0)\setminus D_{e^{-2}}(0))}\right).
\end{aligned}$$

Thus, if $\alpha_0 - 1 > 0$ is small enough, we obtain

$$\|\nabla\Phi_{\alpha_n}\|_{W^{1,2}(D_{e^{5/4}}(0)\setminus D_{e^{-5/4}}(0))} \le C(N)\|\nabla\Phi_{\alpha_n}\|_{L^2(D_{e^2}(0)\setminus D_{e^{-2}}(0))}. \qquad (2.11)$$

According to Sobolev embedding theorem, we know $\nabla\Phi_{\alpha_n} \in L^p(D_{e^{5/4}}(0)\setminus D_{e^{-5/4}}(0))$ for any $p > 1$. Then similarly to deriving (2.11), we can control the norm

$$\|\nabla\Phi_{\alpha_n}\|_{W^{1,p}(D_{e^{9/8}}(0)\setminus D_{e^{-9/8}}(0))} \le C(p, N)\|\nabla\Phi_{\alpha_n}\|_{L^2(D_{e^2}(0)\setminus D_{e^{-2}}(0))}$$

which implies

$$\|\nabla\Phi_{\alpha_n}\|_{C^{1/2}(D_{e^{9/8}}(0)\setminus D_{e^{-9/8}}(0))} \leq C(N)\|\nabla\Phi_{\alpha_n}\|_{L^2(D_{e^2}(0)\setminus D_{e^{-2}}(0))}.$$

To estimate $\|\nabla^2\Phi_{\alpha_n}\|_{C^0}$, noting that Φ_{α_n} satisfies the following elliptic equations that

$$\begin{aligned}\Delta\Phi_{\alpha_n} + 2(\alpha_n - 1)\frac{\frac{r^2}{\rho_n^2}\frac{\partial\Phi_{\alpha_n}}{\partial x^\beta}\frac{\partial\Phi_{\alpha_n}}{\partial x^\gamma}}{1+\frac{r^2}{\rho_n^2}|\nabla\Phi_{\alpha_n}|^2}\frac{\partial^2\Phi_{\alpha_n}}{\partial x^\beta\partial x^\gamma} &= A(\nabla\Phi_{\alpha_n},\nabla\Phi_{\alpha_n}) \\ &\quad - (\alpha_n - 1)\frac{\nabla(\frac{r^2}{\rho_n^2})|\nabla\Phi_{\alpha_n}|^2\nabla\Phi_{\alpha_n}}{1+\frac{r^2}{\rho_n^2}|\nabla\Phi_{\alpha_n}|^2} \\ &:= f(\Phi_{\alpha_n},\nabla\Phi_{\alpha_n}),\end{aligned}$$

where

$$\Delta + 2(\alpha_n - 1)\frac{\frac{r^2}{\rho_n^2}\frac{\partial\Phi_{\alpha_n}}{\partial x^\beta}\frac{\partial\Phi_{\alpha_n}}{\partial x^\gamma}}{1+\frac{r^2}{\rho_n^2}|\nabla\Phi_{\alpha_n}|^2}\frac{\partial^2}{\partial x^\beta\partial x^\gamma}$$

is an elliptic operator when n is big enough. Then by the standard Schauder estimate, we get

$$\begin{aligned}&\|\nabla\Phi_{\alpha_n}\|_{C^1(D_e(0)\setminus D_{\frac{1}{e}}(0))} \\ &\quad\leq C\left(\|f\|_{C^{\frac12}(D_{e^{9/8}}(0)\setminus D_{e^{-9/8}}(0))} + \|\nabla\Phi_{\alpha_n}\|_{C^0(D_{e^{9/8}}(0)\setminus D_{e^{-9/8}}(0))}\right) \\ &\quad\leq C(N)\|\nabla\Phi_{\alpha_n}\|_{L^2(D_{e^2}(0)\setminus D_{e^{-2}}(0))}.\end{aligned}$$

By the standard scaling argument, we get

$$\begin{aligned}&\|\,|x|\cdot|\nabla\Phi_{\alpha_n}|\,\|_{C^0(D_{e^{t+1}}(0)\setminus D_{e^{t-1}}(0))} + \|\,|x|^2\cdot\nabla^2|\Phi_{\alpha_n}|\,\|_{C^0(D_{e^{t+1}}(0)\setminus D_{e^{t-1}}(0))} \\ &\quad\leq C(N)\|\nabla\Phi_{\alpha_n}\|_{L^2(D_{e^{t+2}}(0)\setminus D_{e^{t-2}}(0))}.\end{aligned}$$

Then by a direct computation, we have

$$\begin{aligned}\|\nabla\phi_{\alpha_n}\|_{C^0([t-1,t+1]\times S^1)} &= \|\,|x||\nabla\Phi_{\alpha_n}|\,\|_{C^0(D_{e^{t+1}}(0)\setminus D_{e^{t-1}}(0))} \\ &\leq C(N)\|\nabla\phi_{\alpha_n}\|_{L^2([t-2,t+2]\times S^1)}\end{aligned}$$

and

$$\begin{aligned}\|\nabla^2\phi_{\alpha_n}\|_{C^0([t-1,t+1]\times S^1)} &\leq C(\|\,|x|^2|\nabla^2\Phi_{\alpha_n}|\,\|_{C^0(D_{e^{t+1}}(0)\setminus D_{e^{t-1}}(0))} \\ &\quad + \|\,|x||\nabla\Phi_{\alpha_n}|\,\|_{C^0(D_{e^{t+1}}(0)\setminus D_{e^{t-1}}(0))}) \\ &\leq C(N)\|\nabla\Phi_{\alpha_n}\|_{L^2(D_{e^{t+2}}(0)\setminus D_{e^{t-2}}(0))} \\ &= C(N)\|\nabla\phi_{\alpha_n}\|_{L^2([t-2,t+2]\times S^1)},\end{aligned}$$

which imply the conclusion of lemma immediately. □

At the end of this subsection, we prove some three circle lemmas which will be used in our later proof.

Let $L > 0$ be a constant, denote $P_i = [(i-1)L, iL]\times S^1$ and

$$\|u\|^2_{L^2(P_i)} = \int_{P_i}|u|^2 dtd\theta.$$

PROPOSITION 2.6. *Suppose* $u \in C^\infty(P_{i-1} \cup P_i \cup P_{i+1})$ *satisfying the equation*

$$\Delta u = A^1 u + A^2 \nabla u + \frac{1}{2\pi}\int_0^{2\pi} A^3 u + A^4 \nabla u d\theta + f \tag{2.12}$$

with L *is given and large. Then there exists a positive constant* δ_0 *such that if*

$$\max_{i-1,i,i+1} \|f\|_{L^2(P_j)} \leq \delta_0 \|u\|_{L^2(P_i)}$$

and $\|A^j\|_{L^\infty} \leq \delta_0$ *for* $j = 1, \dots, 4$ *and*

$$|\int_{(i-1)L\times S^1} u d\theta| + |\int_{iL\times S^1} u d\theta| \leq \delta_0 \|u\|_{L^2(P_i)}, \tag{2.13}$$

then

(a) $\|u\|_{L^2(P_{i+1})} \leq e^{-\frac{1}{2}L}\|u\|_{L^2(P_i)}$ *implies* $\|u\|_{L^2(P_i)} \leq e^{-\frac{1}{2}L}\|u\|_{L^2(P_{i-1})}$;
(b) $\|u\|_{L^2(P_{i-1})} \leq e^{-\frac{1}{2}L}\|u\|_{L^2(P_i)}$ *implies* $\|u\|_{L^2(P_i)} \leq e^{-\frac{1}{2}L}\|u\|_{L^2(P_{i+1})}$;
(c) *either* $\|u\|_{L^2(P_i)} \leq e^{-\frac{1}{2}L}\|u\|_{L^2(P_{i+1})}$ *or* $\|u\|_{L^2(P_i)} \leq e^{-\frac{1}{2}L}\|u\|_{L^2(P_{i-1})}$.

PROOF. Setting

$$U := u \circ F^{-1},$$

where F was defined by (2.7), from the fact that

$$\Delta U(x) = \frac{1}{|x|^2}\Delta u \circ F^{-1}(x),$$

it is easy to see that $U \in C^\infty(D_{e^{(i+1)L}}(0) \setminus D_{e^{(i-2)L}}(0))$ and satisfies

$$\Delta U = \widetilde{A}^1 U + \widetilde{A}^2 \nabla U + \frac{1}{2\pi}\int_0^{2\pi} \widetilde{A}^3 U + \widetilde{A}^4 \nabla U d\theta + \widetilde{f} \tag{2.14}$$

where

$$\widetilde{A}^1 = \frac{1}{|x|^2}A^1,\ \widetilde{A}^3 = \frac{1}{|x|^2}A^3,$$
$$\widetilde{A}^2 = \frac{1}{|x|^2}A^2\nabla F \circ F^{-1},\ \widetilde{A}^4 = \frac{1}{|x|^2}A^4\nabla F \circ F^{-1},\ \widetilde{f} = \frac{1}{|x|^2}f.$$

Thus, there hold

$$\max_{i-1,i,i+1}\left(\int_{D_{e^{iL}}(0)\setminus D_{e^{(i-1)L}}(0)} |x|^2|\widetilde{f}|^2 dx\right)^{\frac{1}{2}} \leq \delta_0 \left(\int_{D_{e^{iL}}(0)\setminus D_{e^{(i-1)L}}(0)} \frac{1}{|x|^2}|U|^2 dx\right)^{\frac{1}{2}}$$

and

$$|x|^2\left(|\widetilde{A}^1| + |\widetilde{A}^3|\right) + |x|\left(|\widetilde{A}^2| + |\widetilde{A}^4|\right) \leq C\delta_0$$

and

$$\left|\int_0^{2\pi} U(e^{(i-1)L},\theta)d\theta\right| + \left|\int_0^{2\pi} U(e^{iL},\theta)d\theta\right| \leq \delta_0 \|\frac{U}{|x|}\|_{L^2(D_{e^{iL}}(0)\setminus D_{e^{(i-1)L}}(0))}. \tag{2.15}$$

Then the conclusions of this proposition follow immediately from Theorem 3.3 in [**11**]. One can also refer to Proposition 2.1 in [**22**]. □

For equation (2.12), with the help of (2.14) and by Lemma 3.2 in [**11**], we have the following interior L^2-estimate.

LEMMA 2.7. *Suppose $u \in W^{2,2}([-2,2]\times S^1)$ satisfying the equation*

$$\Delta u = A^1 u + A^2 \nabla u + \frac{1}{2\pi}\int_0^{2\pi} A^3 u + A^4 \nabla u d\theta + f$$

where

$$\|A^i\|_{L^\infty([-2,2]\times S^1)} \le \delta_1,\ i = 1,\dots,4 \ \text{ and } \ \|f\|_{L^2([-2,2]\times S^1)} \le C.$$

Then there exists a positive constant $\delta_0 > 0$ such that if $\delta_1 \le \delta_0$, there holds

$$\|u\|_{W^{2,2}([-1,1]\times S^1)} \le C(\|u\|_{L^2([-2,2]\times S^1)} + \|f\|_{L^2([-2,2]\times S^1)}).$$

Here, we want to remark that in Lemma 3.2 of [**11**], there is a smallness assumption on the L^4 norm of the coefficients $\|A^i\|_{L^4}$. But if we assume a better regularity on the coefficients, for example L^∞, then the smallness assumption in the above lemma is not necessary, i.e., we just need to assume $\|A^i\|_{L^\infty} \le C$. One can refer to a similar proof as in [**16**] and we leave it to interested readers.

With the help of (2.14), by Lemma 3.4 and Corollary 3.5 in [**11**], we have the following decay estimate:

LEMMA 2.8. *Let δ_0, $L > 0$ be the constant in Proposition 2.6. Let $T > 0$ and we also denote $P_i = [-T + (i-1)L, -T + iL] \times S^1$. Suppose $u \in C^\infty((-T, -T + lL) \times S^1)$ for some positive integer l, satisfying the equation (2.12) and assuming that $\|A^j\|_{L^\infty} \le \delta_0$ for $j = 1,\dots,4$ and*

$$\int_{\{t\}\times S^1} u d\theta = 0,\ \ -T < t < -T + lL. \tag{2.16}$$

Then for any $1 \le i \le l$, there holds

$$\|\nabla u\|_{L^2(P_i)} \le C \max_{j=1,\dots,l} \|f\|_{L^2(P_j)} + C(\|u\|_{L^2(P_1)} + \|u\|_{L^2(P_l)}) \left(e^{-\frac{(l-i)L}{2}} + e^{-\frac{iL}{2}}\right).$$

2.3. Generalized energy identity

In this subsection, we will prove our main Theorem 2.1 about a generalized energy identity on the first type neck. The idea is first to estimate the energy of tangential part by using the method of integrating by parts. Then with the help of Pohozaev type constant, we can control the energy of the radial part. The key point here is to deal with some additional terms appearing in the formula of Pohozaev type constant.

We start by showing that there is no energy concentration in the energy of tangential part.

LEMMA 2.9. *Let ϕ_{α_n} be a sequence of α_n-harmonic maps from (Q_n, g_n) to N with uniformly bounded α-energy*

$$\int_{Q_n} (1 + |\nabla_{g_n}\phi_{\alpha_n}|^2)^{\alpha_n} dvol_{g_n} \le \Lambda < \infty,$$

which satisfies

$$\lim_{n\to\infty} \sup_{t\in[T_n^1, T_n^2]} E(\phi_{\alpha_n}, [t-1, t+1]) = 0, \tag{2.17}$$

where $-T_n + 2 \le T_n^1 < T_n^2 \le T_n - 2$. Then there holds

$$\lim_{n\to\infty} \int_{[T_n^1, T_n^2]\times S^1} \left|\frac{\partial \phi_{\alpha_n}}{\partial \theta}\right|^2 dt d\theta = 0.$$

PROOF. By (2.2), we have

$$
\begin{aligned}
\Delta\phi_{\alpha_n} &= -A(\nabla\phi_{\alpha_n},\nabla\phi_{\alpha_n}) - (\alpha_n-1)\frac{\nabla|\nabla_{g_n}\phi_{\alpha_n}|^2\nabla\phi_{\alpha_n}}{1+|\nabla_{g_n}\phi_{\alpha_n}|^2}\\
&= -A(\nabla\phi_{\alpha_n},\nabla\phi_{\alpha_n}) - (\alpha_n-1)\frac{\nabla(\frac{1}{\rho_n^2}|\nabla\phi_{\alpha_n}|^2)\nabla\phi_{\alpha_n}}{1+\frac{1}{\rho_n^2}|\nabla\phi_{\alpha_n}|^2}\\
&= -A(\nabla\phi_{\alpha_n},\nabla\phi_{\alpha_n}) - (\alpha_n-1)\frac{\frac{1}{\rho_n^2}\nabla|\nabla\phi_{\alpha_n}|^2\nabla\phi_{\alpha_n}}{1+\frac{1}{\rho_n^2}|\nabla\phi_{\alpha_n}|^2}\\
&\quad - (\alpha_n-1)\frac{\nabla(\frac{1}{\rho_n^2})|\nabla\phi_{\alpha_n}|^2\nabla\phi_{\alpha_n}}{1+\frac{1}{\rho_n^2}|\nabla\phi_{\alpha_n}|^2},
\end{aligned}\tag{2.18}
$$

where

$$
\left|(\alpha_n-1)\frac{\frac{1}{\rho_n^2}\nabla|\nabla\phi_{\alpha_n}|^2\nabla\phi_{\alpha_n}}{1+\frac{1}{\rho_n^2}|\nabla\phi_{\alpha_n}|^2}\right| \le C(\alpha_n-1)|\nabla^2\phi_{\alpha_n}|\tag{2.19}
$$

and

$$
\begin{aligned}
\left|(\alpha_n-1)\frac{\nabla(\frac{1}{\rho_n^2})|\nabla\phi_{\alpha_n}|^2\nabla\phi_{\alpha_n}}{1+\frac{1}{\rho_n^2}|\nabla\phi_{\alpha_n}|^2}\right| &= \left|(\alpha_n-1)\frac{\frac{2}{\rho_n}\sin(\frac{l_nt}{2\pi})|\nabla\phi_{\alpha_n}|^2\frac{\partial\phi_{\alpha_n}}{\partial t}}{1+\frac{1}{\rho_n^2}|\nabla\phi_{\alpha_n}|^2}\right|\\
&\le C(\alpha_n-1)\left|\rho_n\sin(\frac{l_nt}{2\pi})\right|\cdot\left|\frac{\partial\phi_{\alpha_n}}{\partial t}\right|\\
&\le C(\alpha_n-1)\left|\frac{\partial\phi_{\alpha_n}}{\partial t}\right|
\end{aligned}\tag{2.20}
$$

where we used the fact that

$$
\frac{d\rho_n}{dt} = \rho_n^2\sin(\frac{l_nt}{2\pi}),\quad |\rho_n(t)|\le C
$$

for all $t\in[-T_n,T_n]$.

Denoting

$$
\phi^*_{\alpha_n}(t) = \frac{1}{2\pi}\int_0^{2\pi}\phi_{\alpha_n}(t,\theta)d\theta,\tag{2.21}
$$

then by (2.17) and Lemma 2.5, we know that for any $\epsilon>0$, there holds

$$
|\phi_{\alpha_n}(t,\theta)-\phi^*_{\alpha_n}(t)| \le C|\nabla\phi_{\alpha_n}(t,\theta)| \le CE(\phi_{\alpha_n},[t-1,t+1]\times S^1)\le C\epsilon,\tag{2.22}
$$

when n is big enough.

Integrating by parts and using equation (2.18), we get

$$\begin{aligned}
&\int_{[T_n^1,T_n^2]\times S^1}\nabla\phi_{\alpha_n}\nabla(\phi_{\alpha_n}-\phi^*_{\alpha_n})dtd\theta-\int_{\{T_n^2\}\times S^1}\frac{\partial\phi_{\alpha_n}}{\partial t}(\phi_{\alpha_n}-\phi^*_{\alpha_n})d\theta\\
&\quad+\int_{\{T_n^1\}\times S^1}\frac{\partial\phi_{\alpha_n}}{\partial t}(\phi_{\alpha_n}-\phi^*_{\alpha_n})d\theta\\
&=\int_{[T_n^1,T_n^2]\times S^1}-\Delta\phi_{\alpha_n}(\phi_{\alpha_n}-\phi^*_{\alpha_n})dtd\theta\\
&=\int_{[T_n^1,T_n^2]\times S^1}A(\nabla\phi_{\alpha_n},\nabla\phi_{\alpha_n})(\phi_{\alpha_n}-\phi^*_{\alpha_n})dtd\theta\\
&\quad+\int_{[T_n^1,T_n^2]\times S^1}\left((\alpha_n-1)\frac{\frac{1}{\rho_n^2}\nabla|\nabla\phi_{\alpha_n}|^2\nabla\phi_{\alpha_n}}{1+\frac{1}{\rho_n^2}|\nabla\phi_{\alpha_n}|^2}+(\alpha_n-1)\frac{\nabla(\frac{1}{\rho_n^2})|\nabla\phi_{\alpha_n}|^2\nabla\phi_{\alpha_n}}{1+\frac{1}{\rho_n^2}|\nabla\phi_{\alpha_n}|^2}\right)\\
&\quad\times(\phi_{\alpha_n}-\phi^*_{\alpha_n})dtd\theta\\
&\le C(N)\epsilon\int_{[T_n^1,T_n^2]\times S^1}|\nabla\phi_{\alpha_n}|^2dtd\theta\\
&\quad+C(N)(\alpha_n-1)\int_{[T_n^1,T_n^2]\times S^1}(|\nabla^2\phi_{\alpha_n}|+|\nabla\phi_{\alpha_n}|)|\phi_{\alpha_n}-\phi^*_{\alpha_n}|dtd\theta.
\end{aligned}$$

On the one hand, noting that

$$\begin{aligned}
&\int_{[T_n^1,T_n^2]\times S^1}\nabla\phi_{\alpha_n}\nabla\phi^*_{\alpha_n}dtd\theta\\
&\quad=\int_{[T_n^1,T_n^2]\times S^1}\frac{\partial\phi_{\alpha_n}}{\partial t}\frac{\partial\phi^*_{\alpha_n}}{\partial t}dtd\theta\\
&\quad\le\left(\int_{[T_n^1,T_n^2]\times S^1}|\frac{\partial\phi_{\alpha_n}}{\partial t}|^2dtd\theta\right)^{\frac12}\left(\int_{[T_n^1,T_n^2]\times S^1}|\frac{\partial\phi^*_{\alpha_n}}{\partial t}|^2dtd\theta\right)^{\frac12}\\
&\quad\le\int_{[T_n^1,T_n^2]\times S^1}|\frac{\partial\phi_{\alpha_n}}{\partial t}|^2dtd\theta,
\end{aligned}$$

we obtain

$$\begin{aligned}
&\int_{[T_n^1,T_n^2]\times S^1}\nabla\phi_{\alpha_n}\nabla(\phi_{\alpha_n}-\phi^*_{\alpha_n})dtd\theta\\
&\quad\ge\int_{[T_n^1,T_n^2]\times S^1}|\nabla\phi_{\alpha_n}|^2dtd\theta-\int_{[T_n^1,T_n^2]\times S^1}|\frac{\partial\phi_{\alpha_n}}{\partial t}|^2dtd\theta\\
&\quad=\int_{[T_n^1,T_n^2]\times S^1}|\frac{\partial\phi_{\alpha_n}}{\partial\theta}|^2dtd\theta.
\end{aligned}$$

On the other hand, by Lemma 2.5, we have

$$\begin{aligned}
\int_{[T_n^1,T_n^2]\times S^1}|\nabla^2\phi_{\alpha_n}|^2dtd\theta&\le\sum_{i=1}^{[T_n^2-T_n^1]+1}\int_{[T_n^1+(i-1),T_n^1+i]\times S^1}|\nabla^2\phi_{\alpha_n}|^2dtd\theta\\
&\le C\sum_{i=1}^{[T_n^2-T_n^1]+1}\int_{[T_n^1+(i-2),T_n^2+(i+1)]\times S^1}|\nabla\phi_{\alpha_n}|^2dtd\theta\\
&\le CE(\phi_{\alpha_n},[T_n^1-1,T_n^2+2]\times S^1)\le C\Lambda,
\end{aligned}$$

then it yields

$$\begin{aligned}&\int_{[T_n^1,T_n^2]\times S^1}(|\nabla^2\phi_{\alpha_n}|+|\nabla\phi_{\alpha_n}|)|\phi_{\alpha_n}-\phi^*_{\alpha_n}|dtd\theta\\&\le\left(\int_{[T_n^1,T_n^2]\times S^1}|\nabla^2\phi_{\alpha_n}|^2dtd\theta\right)^{\frac12}\left(\int_{[T_n^1,T_n^2]\times S^1}|\phi_{\alpha_n}-\phi^*_{\alpha_n}|^2dtd\theta\right)^{\frac12}\\&\quad+\left(\int_{[T_n^1,T_n^2]\times S^1}|\nabla\phi_{\alpha_n}|^2dtd\theta\right)^{\frac12}\left(\int_{[T_n^1,T_n^2]\times S^1}|\phi_{\alpha_n}-\phi^*_{\alpha_n}|^2dtd\theta\right)^{\frac12}\\&\le C\sqrt{\Lambda}\left(\int_{[T_n^1,T_n^2]\times S^1}|\nabla\phi_{\alpha_n}|^2dtd\theta\right)^{\frac12}\le C\Lambda,\end{aligned}$$

where we used Poincaré inequality.

Combining these together, we arrived at

$$\begin{aligned}\int_{[T_n^1,T_n^2]\times S^1}|\frac{\partial\phi_{\alpha_n}}{\partial\theta}|^2dtd\theta\le C\left[\epsilon+(\alpha_n-1)\right]+\left|\int_{\{T_n^2\}\times S^1}\frac{\partial\phi_{\alpha_n}}{\partial t}(\phi_{\alpha_n}-\phi^*_{\alpha_n})d\theta\right|\\+\left|\int_{\{T_n^1\}\times S^1}\frac{\partial\phi_{\alpha_n}}{\partial t}(\phi_{\alpha_n}-\phi^*_{\alpha_n})d\theta\right|.\end{aligned}$$

For the boundary term, by trace theory and (2.22), we have

$$\begin{aligned}\left|\int_{\{T_n^2\}\times S^1}\frac{\partial\phi_{\alpha_n}}{\partial t}(\phi_{\alpha_n}-\phi^*_{\alpha_n})d\theta\right|&\le C\epsilon\int_{\{T_n^2\}\times S^1}|\frac{\partial\phi_{\alpha_n}}{\partial t}|d\theta\\&\le C\epsilon\|\nabla\phi_{\alpha_n}\|_{W^{1,2}([T_n^2-1,T_n^2+1]\times S^1)}\\&\le C\epsilon\|\nabla\phi_{\alpha_n}\|_{L^2([T_n^2-2,T_n^2+2]\times S^1)}\le C\epsilon.\end{aligned}$$

Similarly, there holds

$$\left|\int_{\{T_n^1\}\times S^1}\frac{\partial\phi_{\alpha_n}}{\partial t}(\phi_{\alpha_n}-\phi^*_{\alpha_n})d\theta\right|\le C\epsilon.$$

Thus, we conclude that

$$\int_{[T_n^1,T_n^2]\times S^1}|\frac{\partial\phi_{\alpha_n}}{\partial\theta}|^2dtd\theta\le C\left[\epsilon+(\alpha_n-1)\right],$$

which implies the conclusion of lemma. □

As a corollary of above lemma, we have

COROLLARY 2.10. *Under the assumption of Lemma 2.9, if*

$$\limsup_{n\to\infty}(l_n)^{2(1-\alpha_n)}\le C<\infty,$$

then we have

$$\lim_{n\to\infty}\int_{[T_n^1,T_n^2]\times S^1}(1+|\nabla_{g_n}\phi_{\alpha_n}|^2)^{\alpha_n-1}|\frac{\partial\phi_{\alpha_n}}{\partial\theta}|^2dtd\theta=0. \tag{2.23}$$

PROOF. Noting that (2.17) holds, Lemma 2.5 tells us that

$$|\nabla\phi_{\alpha_n}|(t,\theta) \le CE(\phi_{\alpha_n},[t-1,t+1]\times S^1),\ (t,\theta)\in[T_n^1,T_n^2]\times S^1, \tag{2.24}$$

where $C>0$ is independent of n, which implies

$$\int_{[T_n^1,T_n^2]\times S^1}(1+|\nabla_{g_n}\phi_{\alpha_n}|^2)^{\alpha_n-1}|\frac{\partial\phi_{\alpha_n}}{\partial\theta}|^2dtd\theta \le Cl_n^{2-2\alpha_n}\int_{[T_n^1,T_n^2]\times S^1}|\frac{\partial\phi_{\alpha_n}}{\partial\theta}|^2dtd\theta.$$

Then the conclusion of corollary follows from Lemma 2.9 immediately. □

At the end of this subsection, we give the proof of Theorem 2.1.

Proof of Theorem 2.1. We divide the proof into following 4 cases:

CASE 1. $\lim_{n\to\infty}\big(T_n^1-(-T_n)\big)=\infty$ and $\lim_{n\to\infty}(T_n-T_n^2)=\infty$.

By Proposition 2.4, we have

$$\begin{aligned}
&(1-\frac{1}{2\alpha_n})\int_{[T_n^1,T_n^2]\times S^1}(1+|\nabla_{g_n}\phi_{\alpha_n}|^2)^{\alpha_n-1}|\frac{\partial\phi_{\alpha_n}}{\partial t}|^2dtd\theta\\
&\quad-\frac{1}{2\alpha_n}\int_{[T_n^1,T_n^2]\times S^1}(1+|\nabla_{g_n}\phi_{\alpha_n}|^2)^{\alpha_n-1}|\frac{\partial\phi_{\alpha_n}}{\partial\theta}|^2dtd\theta\\
&=\frac{1}{2\alpha_n}\int_{[T_n^1,T_n^2]\times S^1}(1+|\nabla_{g_n}\phi_{\alpha_n}|^2)^{\alpha_n-1}\rho_n^2dtd\theta\\
&\quad-\frac{1}{\alpha_n}\int_{T_n^1}^{T_n^2}\int_{[0,s]\times S^1}(1+|\nabla_{g_n}\phi_{\alpha_n}|^2)^{\alpha_n-1}\rho_n^3\sin(\frac{l_nt}{2\pi})dtd\theta ds\\
&\quad+(1-\frac{1}{\alpha_n})\int_{T_n^1}^{T_n^2}\int_{[0,s]\times S^1}(1+|\nabla_{g_n}\phi_{\alpha_n}|^2)^{\alpha_n-1}|\nabla\phi_{\alpha_n}|^2\rho_n\sin(\frac{l_nt}{2\pi})dtd\theta ds\\
&\quad+\kappa_n(\phi_{\alpha_n})(T_n^2-T_n^1)\\
&=\mathbb{I}+\mathbb{II}+\mathbb{III}+\kappa_n(\phi_{\alpha_n})(T_n^2-T_n^1).
\end{aligned}$$

Next, we estimate $\mathbb{I}$, $\mathbb{II}$, $\mathbb{III}$ separately. Firstly, by Young's inequality, we have

$$\begin{aligned}
\lim_{n\to\infty}\mathbb{I}&=\lim_{n\to\infty}\frac{1}{2\alpha_n}\int_{[T_n^1,T_n^2]\times S^1}(1+|\nabla_{g_n}\phi_{\alpha_n}|^2)^{\alpha_n-1}\rho_n^2dtd\theta\\
&\le\lim_{R\to\infty}\lim_{n\to\infty}C\int_{[-T_n+R,T_n-R]\times S^1}(1+|\nabla_{g_n}\phi_{\alpha_n}|^2)^{\alpha_n-1}\rho_n^2dtd\theta\\
&\le C\lim_{R\to\infty}\lim_{n\to\infty}\left(\int_{[-T_n+R,T_n-R]\times S^1}(1+|\nabla_{g_n}\phi_{\alpha_n}|^2)^{\alpha_n}\rho_n^2dtd\theta\right)^{1-\frac{1}{\alpha_n}}\\
&\quad\times\left(\int_{[-T_n+R,T_n-R]\times S^1}\rho_n^2dtd\theta\right)^{\frac{1}{\alpha_n}}\\
&\le C\Lambda\lim_{R\to\infty}\lim_{n\to\infty}\left(\int_{[-T_n+R,T_n-R]\times S^1}\rho_n^2dtd\theta\right)^{\frac{1}{\alpha_n}}\\
&=C\Lambda\lim_{R\to\infty}\lim_{n\to\infty}\left(2l_n\tan\frac{l_nt}{2\pi}\Big|_0^{T_n-R}\right)^{\frac{1}{\alpha_n}}\le\lim_{R\to\infty}\frac{C}{R}=0,
\end{aligned}\tag{2.25}$$

where we used

$$\lim_{n\to\infty} \frac{\tan(\frac{l_n}{2\pi}(T_n - R))}{\frac{2}{l_n(1+\frac{R}{\pi})}} = 1.$$

For the second term $\mathbb{III}$, it is easy to see that

$$\begin{aligned}
\lim_{n\to\infty} |\mathbb{III}| &\le \lim_{R\to\infty}\lim_{n\to\infty}\left|\frac{1}{\alpha_n}\int_{-T_n+R}^{T_n-R}\int_{[0,s]\times S^1}(1+|\nabla_{g_n}\phi_{\alpha_n}|^2)^{\alpha_n-1}\rho_n^3\sin(\frac{l_n t}{2\pi})dtd\theta ds\right| \\
&\le \lim_{R\to\infty}\lim_{n\to\infty}\left|\frac{1}{\alpha_n}\int_{-T_n+R}^{0}\int_{[s,0]\times S^1}(1+|\nabla_{g_n}\phi_{\alpha_n}|^2)^{\alpha_n-1}\rho_n^3\sin(\frac{l_n t}{2\pi})dtd\theta ds\right| \\
&\quad + \lim_{R\to\infty}\lim_{n\to\infty}\left|\frac{1}{\alpha_n}\int_{0}^{T_n-R}\int_{[0,s]\times S^1}(1+|\nabla_{g_n}\phi_{\alpha_n}|^2)^{\alpha_n-1}\rho_n^3\sin(\frac{l_n t}{2\pi})dtd\theta ds\right|.
\end{aligned}$$

Now, we claim

$$(T_n - R - t)\rho_n \sin(\frac{l_n t}{2\pi}) \le 2,\ t \in [0, T_n - R]. \tag{2.26}$$

In fact, denoting

$$f(t) := (T_n - R - t)l_n \tan(\frac{l_n t}{2\pi}),\ t \in [0, T_n - R],$$

it is easy to see that $f(0) = f(T_n - R) = 0$ and $f(t) \ge 0$. So there exists $t_n \in (0, T_n - R)$ such that

$$f(t_n) = \max_{0\le t\le T_n - R} f(t).$$

Moreover, we have

$$\begin{aligned}
0 = f'(t_n) &= -l_n \tan(\frac{l_n t_n}{2\pi}) + (T_n - R - t_n)l_n \sec^2(\frac{l_n t}{2\pi})\frac{l_n}{2\pi} \\
&= l_n\left(-\tan(\frac{l_n t_n}{2\pi}) + (T_n - R - t_n)\frac{l_n}{2\pi}(1+\tan^2(\frac{l_n t_n}{2\pi}))\right)
\end{aligned}$$

which implies

$$\tan(\frac{l_n t_n}{2\pi}) = \frac{1 \pm \sqrt{1-4c_n^2}}{2c_n}$$

where $c_n = (T_n - R - t_n)\frac{l_n}{2\pi}$.

Therefore, we have

$$f(t) \le f(t_n) = (T_n - R - t_n)l_n \tan(\frac{l_n t_n}{2\pi}) = 2\pi c_n \frac{1 \pm \sqrt{1-4c_n^2}}{2c_n} \le 4\pi,$$

which implies (2.26) immediately.

By a direct computation, we get

$$
\begin{aligned}
&\lim_{R\to\infty}\lim_{n\to\infty}\frac{1}{\alpha_n}\int_0^{T_n-R}\int_{[0,s]\times S^1}(1+|\nabla_{g_n}\phi_{\alpha_n}|^2)^{\alpha_n-1}\rho_n^3\sin(\frac{l_nt}{2\pi})dtd\theta ds\\
&=\lim_{R\to\infty}\lim_{n\to\infty}\frac{1}{\alpha_n}\int_{[0,T_n-R]\times S^1}\int_t^{T_n-R}ds(1+|\nabla_{g_n}\phi_{\alpha_n}|^2)^{\alpha_n-1}\rho_n^3\sin(\frac{l_nt}{2\pi})dtd\theta\\
&=\lim_{R\to\infty}\lim_{n\to\infty}\frac{1}{\alpha_n}\int_{[0,T_n-R]\times S^1}(1+|\nabla_{g_n}\phi_{\alpha_n}|^2)^{\alpha_n-1}(T_n-R-t)\rho_n^3\sin(\frac{l_nt}{2\pi})dtd\theta\\
&\le C\lim_{R\to\infty}\lim_{n\to\infty}\int_{[0,T_n-R]\times S^1}(1+|\nabla_{g_n}\phi_{\alpha_n}|^2)^{\alpha_n-1}\rho_n^2dtd\theta=0,
\end{aligned}
\tag{2.27}
$$

where we used (2.26) and (2.25).

Similarly,

$$
\lim_{R\to\infty}\lim_{n\to\infty}\frac{1}{\alpha_n}\int_{-T_n+R}^{0}\int_{[s,0]\times S^1}(1+|\nabla_{g_n}\phi_{\alpha_n}|^2)^{\alpha_n-1}\rho_n^3\sin(\frac{l_nt}{2\pi})dtd\theta ds=0.
$$

Thus, we get

$$
\lim_{n\to\infty}\mathbb{II}=0.
$$

For the third term $\mathbb{III}$, we have

$$
\begin{aligned}
\lim_{n\to\infty}|\mathbb{III}|\le&\lim_{R\to\infty}\lim_{n\to\infty}\Big|(1-\frac{1}{\alpha_n})\int_{-T_n+R}^{T_n-R}\int_{[0,s]\times S^1}(1+|\nabla_{g_n}\phi_{\alpha_n}|^2)^{\alpha_n-1}|\nabla\phi_{\alpha_n}|^2\rho_n\\
&\sin(\frac{l_nt}{2\pi})dtd\theta ds\Big|\\
=&\lim_{R\to\infty}\lim_{n\to\infty}\Big|(1-\frac{1}{\alpha_n})\int_{-T_n+R}^{0}\int_{[s,0]\times S^1}(1+|\nabla_{g_n}\phi_{\alpha_n}|^2)^{\alpha_n-1}|\nabla\phi_{\alpha_n}|^2\rho_n\\
&\sin(\frac{l_nt}{2\pi})dtd\theta ds\Big|\\
&+\lim_{R\to\infty}\lim_{n\to\infty}\Big|(1-\frac{1}{\alpha_n})\int_{0}^{T_n-R}\int_{[0,s]\times S^1}(1+|\nabla_{g_n}\phi_{\alpha_n}|^2)^{\alpha_n-1}|\nabla\phi_{\alpha_n}|^2\rho_n\\
&\sin(\frac{l_nt}{2\pi})dtd\theta ds\Big|.
\end{aligned}
$$

Computing directly and using (2.26), we obtain

$$
\begin{aligned}
&(1-\frac{1}{\alpha_n})\int_0^{T_n-R}\int_{[0,s]\times S^1}(1+|\nabla_{g_n}\phi_{\alpha_n}|^2)^{\alpha_n-1}|\nabla\phi_{\alpha_n}|^2\rho_n\sin(\frac{l_nt}{2\pi})dtd\theta ds\\
&=(1-\frac{1}{\alpha_n})\int_{[0,T_n-R]\times S^1}\int_t^{T_n-R}(1+|\nabla_{g_n}\phi_{\alpha_n}|^2)^{\alpha_n-1}|\nabla\phi_{\alpha_n}|^2\rho_n\sin(\frac{l_nt}{2\pi})dsdtd\theta\\
&=(1-\frac{1}{\alpha_n})\int_{[0,T_n-R]\times S^1}(1+|\nabla_{g_n}\phi_{\alpha_n}|^2)^{\alpha_n-1}|\nabla\phi_{\alpha_n}|^2(T_n-R-t)\frac{l_n}{2\pi}\tan(\frac{l_nt}{2\pi})dtd\theta\\
&\le\frac{1}{2\pi}\max_{0\le t\le T_n-R}\left\{(T_n-R-t)l_n\tan(\frac{l_nt}{2\pi})\right\}(1-\frac{1}{\alpha_n})\\
&\quad\times\int_{[0,T_n-R]\times S^1}(1+|\nabla_{g_n}\phi_{\alpha_n}|^2)^{\alpha_n-1}|\nabla\phi_{\alpha_n}|^2dtd\theta\\
&\le C\Lambda(\alpha_n-1).
\end{aligned}
$$

Similarly, there holds

$$
\begin{aligned}
&\left|(1-\frac{1}{\alpha_n})\int_{-T_n+R}^0\int_{[s,0]\times S^1}(1+|\nabla_{g_n}\phi_{\alpha_n}|^2)^{\alpha_n-1}|\nabla\phi_{\alpha_n}|^2\rho_n\sin(\frac{l_nt}{2\pi})dtd\theta ds\right|\\
&\quad\le C\Lambda(\alpha_n-1).
\end{aligned}
$$

Thus, we get the estimate

$$\mathbb{IIII}\le C\Lambda(\alpha_n-1). \tag{2.28}$$

Combining these together, we arrived at

$$
\begin{aligned}
&\lim_{n\to\infty}\int_{[T_n^1,T_n^2]\times S^1}(1+|\nabla_{g_n}\phi_{\alpha_n}|^2)^{\alpha_n-1}|\frac{\partial\phi_{\alpha_n}}{\partial t}|^2dtd\theta\\
&=\lim_{n\to\infty}\int_{[T_n^1,T_n^2]\times S^1}(1+|\nabla_{g_n}\phi_{\alpha_n}|^2)^{\alpha_n-1}|\frac{\partial\phi_{\alpha_n}}{\partial\theta}|^2dtd\theta+\lim_{n\to\infty}2\kappa_n(\phi_{\alpha_n})(T_n^2-T_n^1)\\
&=\lim_{n\to\infty}2\kappa_n(\phi_{\alpha_n})(T_n^2-T_n^1),
\end{aligned}\tag{2.29}
$$

where the last equality follows from Corollary 2.10.

By (2.25), (2.29), Corollary 2.10 and the fact that

$$\lim_{n\to\infty}\int_{[T_n^1,T_n^2]\times S^1}\rho_n^2dtd\theta\le\lim_{R\to\infty}\lim_{n\to\infty}\int_{[-T_n+R,T_n-R]\times S^1}\rho_n^2dtd\theta=0,$$

we have

$$
\begin{aligned}
&\lim_{n\to\infty}\int_{[T_n^1,T_n^2]\times S^1}\left((1+|\nabla_{g_n}\phi_{\alpha_n}|^2)^{\alpha_n}-1\right)\rho_n^2dtd\theta\\
&=\lim_{n\to\infty}\int_{[T_n^1,T_n^2]\times S^1}\left((1+|\nabla_{g_n}\phi_{\alpha_n}|^2)^{\alpha_n-1}-1\right)\rho_n^2dtd\theta\\
&\quad+\lim_{n\to\infty}\int_{[T_n^1,T_n^2]\times S^1}(1+|\nabla_{g_n}\phi_{\alpha_n}|^2)^{\alpha_n-1}|\nabla\phi_{\alpha_n}|^2dtd\theta\\
&=\lim_{n\to\infty}2\kappa_n(\phi_{\alpha_n})(T_n^2-T_n^1).
\end{aligned}\tag{2.30}
$$

CASE 2. $\lim_{n\to\infty}\big(T_n^1-(-T_n)\big)<\infty$ and $\lim_{n\to\infty}(T_n-T_n^2)=\infty$.

In this case, without loss of generality, we assume $T_n^1=-T_n+2$. Firstly, by the proof of **Case 1**, we get

$$\begin{aligned}&\lim_{R\to\infty}\lim_{n\to\infty}\int_{[T_n^1+R,T_n^2]\times S^1}(1+|\nabla_{g_n}\phi_{\alpha_n}|^2)^{\alpha_n-1}|\nabla\phi_{\alpha_n}|^2dtd\theta\\&=\lim_{n\to\infty}2\kappa_n(\phi_{\alpha_n})(T_n^2-T_n^1).\end{aligned}$$

Secondly, noting that

$$\frac{C}{R}\le\rho_n\le C,\ t\in[-T_n,-T_n+2R]$$

and (2.24), we have

$$\sup_{t\in[-T_n,-T_n+2R]}(1+|\nabla_{g_n}\phi_{\alpha_n}|^2)^{\alpha_n-1}=1+o(n),\tag{2.31}$$

as $n\to\infty$ and

$$\begin{aligned}&\lim_{R\to\infty}\lim_{n\to\infty}\int_{[T_n^1,T_n^1+R]\times S^1}(1+|\nabla_{g_n}\phi_{\alpha_n}|^2)^{\alpha_n-1}|\nabla\phi_{\alpha_n}|^2dtd\theta\\&=\lim_{R\to\infty}\lim_{n\to\infty}\int_{[T_n^1,T_n^1+R]\times S^1}|\nabla\phi_{\alpha_n}|^2dtd\theta=0.\end{aligned}$$

Therefore, we obtain

(2.32)

$$\lim_{n\to\infty}\int_{[T_n^1,T_n^2]\times S^1}(1+|\nabla_{g_n}\phi_{\alpha_n}|^2)^{\alpha_n-1}|\nabla\phi_{\alpha_n}|^2dtd\theta=\lim_{n\to\infty}2\kappa_n(\phi_{\alpha_n})(T_n^2-T_n^1).$$

Thirdly, by (2.25) and (2.31), we have

$$\begin{aligned}&\lim_{n\to\infty}\int_{[T_n^1,T_n^2]\times S^1}\left((1+|\nabla_{g_n}\phi_{\alpha_n}|^2)^{\alpha_n-1}-1\right)\rho_n^2dtd\theta\\&=\lim_{R\to\infty}\lim_{n\to\infty}\int_{[T_n^1+R,T_n^2]\times S^1}\left((1+|\nabla_{g_n}\phi_{\alpha_n}|^2)^{\alpha_n-1}-1\right)\rho_n^2dtd\theta\\&\quad+\lim_{R\to\infty}\lim_{n\to\infty}\int_{[T_n^1,T_n^1+R]\times S^1}\left((1+|\nabla_{g_n}\phi_{\alpha_n}|^2)^{\alpha_n-1}-1\right)\rho_n^2dtd\theta\\&\le C\lim_{R\to\infty}\lim_{n\to\infty}\int_{[T_n^1+R,T_n^2]\times S^1}\rho_n^2dtd\theta+\lim_{R\to\infty}\lim_{n\to\infty}\left(o(n)\cdot\int_{[T_n^1,T_n^1+R]\times S^1}\rho_n^2dtd\theta\right)\\&=0.\end{aligned}$$

Similarly to deriving (2.30), we get the conclusion of the theorem.

CASE 3. $\lim_{n\to\infty}\big(T_n^1-(-T_n)\big)=\infty$ and $\lim_{n\to\infty}(T_n-T_n^2)<\infty$.

This case is similar to **Case 2**.

CASE 4. $\lim_{n\to\infty}(T_n^1-(-T_n))<\infty$ and $\lim_{n\to\infty}(T_n-T_n^2)<\infty$.

From the proof of **Case 2** and **Case 3**, it is not hard to see that this case holds. Thus, we finished the proof of this theorem. □

2.4. Refined neck analysis on the first type neck

In this subsection, our goal is to show that the first type neck converges to a geodesic on the target manifold and furthermore derive a length formula of this neck geodesic.

We first derive an exponential decay of tangential energy in the neck domain by using three circle Lemma 2.8.

LEMMA 2.11. *Let ϕ_{α_n} be a sequence of α_n-harmonic maps from (Q_n, g_n) to N with uniformly bounded α-energy*

$$\int_{Q_n} (1 + |\nabla_{g_n}\phi_{\alpha_n}|^2)^{\alpha_n} dvol_{g_n} \leq \Lambda < \infty,$$

which satisfies

$$\lim_{n\to\infty} \sup_{t\in[T_n^1, T_n^2]} E(\phi_{\alpha_n}, [t-1, t+1] \times S^1) = 0,$$

where $-T_n + 2 \leq T_n^1 < T_n^2 \leq T_n - 2$. Denote

$$P_i = [T_n^1 + (i-1)L, T_n^1 + iL] \times S^1,$$

and without loss of generality, we always assume $T_n^2 - T_n^1 = k_n L$ for some positive integer k_n which goes to infinity as $n \to \infty$. Then there holds

$$\left(\int_{P_i} |\frac{\partial\phi_{\alpha_n}}{\partial\theta}|^2 dt d\theta\right)^{\frac{1}{2}} \leq \left((\alpha_n - 1) + e^{-\frac{iL}{2}} + e^{-\frac{(k_n-i)L}{2}}\right) o(n), \tag{2.33}$$

where $o(n) \to 0$ as $n \to \infty$.

PROOF. Denote

$$\phi_{\alpha_n}^*(r) = \frac{1}{2\pi}\int_0^{2\pi} \phi_{\alpha_n}(r,\theta)d\theta, \quad u_n = \phi_{\alpha_n} - \phi_{\alpha_n}^*.$$

Next, we use the same method as in [**16**] to compute the equation for $\phi_{\alpha_n} - \phi_{\alpha_n}^*$. By equation (2.18), we have

$$\begin{aligned}\Delta\phi_{\alpha_n}^* &= \frac{1}{2\pi}\int_0^{2\pi} \Delta\phi_{\alpha_n} d\theta \\ &= \frac{1}{2\pi}\int_0^{2\pi} \left(A(\phi_{\alpha_n})(d\phi_{\alpha_n}, d\phi_{\alpha_n}) - (\alpha_n - 1)\frac{\nabla|\nabla_{g_n}\phi_{\alpha_n}|^2\nabla\phi_{\alpha_n}}{1 + |\nabla_{g_n}\phi_{\alpha_n}|^2}\right) d\theta.\end{aligned}$$

Computing directly, we have

$$\begin{aligned}&\frac{1}{2\pi}\int_0^{2\pi} A(\phi_{\alpha_n})(d\phi_{\alpha_n}, d\phi_{\alpha_n})d\theta \\ &= \frac{1}{2\pi}\int_0^{2\pi} A(\phi_{\alpha_n})(d\phi_{\alpha_n}, d\phi_{\alpha_n}) - A(\phi_{\alpha_n}^*)(d\phi_{\alpha_n}, d\phi_{\alpha_n}) + A(\phi_{\alpha_n}^*)(d\phi_{\alpha_n}, d\phi_{\alpha_n}) \\ &\quad - A(\phi_{\alpha_n}^*)(d\phi_{\alpha_n}^*, d\phi_{\alpha_n}^*) + A(\phi_{\alpha_n}^*)(d\phi_{\alpha_n}^*, d\phi_{\alpha_n}^*)d\theta \\ &= A(\phi_{\alpha_n}^*)(d\phi_{\alpha_n}^*, d\phi_{\alpha_n}^*) + \frac{1}{2\pi}\int_0^{2\pi} A^3(\phi_{\alpha_n} - \phi_{\alpha_n}^*) + A^4\nabla(\phi_{\alpha_n} - \phi_{\alpha_n}^*)d\theta,\end{aligned}$$

where A^i may change from line to line and it just stands for some expression satisfying

$$|A^3| \leq C(N)|\nabla\phi_{\alpha_n}|^2, \quad |A^4| \leq C(N)|\nabla\phi_{\alpha_n}|.$$

Using the same method, we get

$$\begin{aligned}\Delta(\phi_{\alpha_n}-\phi^*_{\alpha_n}) &= A^1(\phi_{\alpha_n}-\phi^*_{\alpha_n})+A^2\nabla(\phi_{\alpha_n}-\phi^*_{\alpha_n})\\ &\quad+\frac{1}{2\pi}\int_0^{2\pi}A^3(\phi_{\alpha_n}-\phi^*_{\alpha_n})+A^4\nabla(\phi_{\alpha_n}-\phi^*_{\alpha_n})d\theta\\ &\quad-(\alpha_n-1)\frac{\nabla|\nabla_{g_n}\phi_{\alpha_n}|^2\nabla\phi_{\alpha_n}}{1+|\nabla_{g_n}\phi_{\alpha_n}|^2}\\ &\quad+\frac{1}{2\pi}\int_0^{2\pi}(\alpha_n-1)\frac{\nabla|\nabla_{g_n}\phi_{\alpha_n}|^2\nabla\phi_{\alpha_n}}{1+|\nabla_{g_n}\phi_{\alpha_n}|^2}d\theta,\end{aligned}\tag{2.34}$$

where

$$|A^1|+|A^3|\le C(N)|\nabla\phi_{\alpha_n}|^2,\quad |A^2|+|A^4|\le C(N)|\nabla\phi_{\alpha_n}|.$$

By (2.24), we have that for any $\epsilon>0$, there holds

$$\|A^i\|_{L^\infty([T_n^1,T_n^2]\times S^1)}\le\epsilon,$$

when n is big enough.

Substituting $u=u_n:=\phi_{\alpha_n}-\phi^*_{\alpha_n}$ and

$$f=-(\alpha_n-1)\frac{\nabla|\nabla_{g_n}\phi_{\alpha_n}|^2\nabla\phi_{\alpha_n}}{1+|\nabla_{g_n}\phi_{\alpha_n}|^2}+\frac{1}{2\pi}\int_0^{2\pi}(\alpha_n-1)\frac{\nabla|\nabla_{g_n}\phi_{\alpha_n}|^2\nabla\phi_{\alpha_n}}{1+|\nabla_{g_n}\phi_{\alpha_n}|^2}d\theta$$

in Lemma 2.8 and noting that

$$\begin{aligned}\|f\|_{L^2(P_i)} &\le C(\alpha_n-1)\left(\|\nabla^2\phi_{\alpha_n}\|_{L^2(P_i)}+\|\nabla\phi_{\alpha_n}\|_{L^2(P_i)}\right)\\ &\le C(\alpha_n-1)\|\nabla\phi_{\alpha_n}\|_{L^2(P_{i-1}\cup P_i\cup P_{i+1})},\\ &\qquad i=1,\dots,k_n,\end{aligned}$$

where we used (2.19), (2.20), Lemma 2.5 and Lemma 2.7, we obtain the energy decay in the θ-direction,

$$\begin{aligned}\|\frac{\partial\phi_{\alpha_n}}{\partial\theta}\|_{L^2(P_i)} &\le\|\nabla(\phi_{\alpha_n}-\phi^*_{\alpha_n})\|_{L^2(P_i)}\\ &\le C(\|\phi_{\alpha_n}-\phi^*_{\alpha_n}\|_{L^2(P_{i-1}\cup P_i\cup P_{i+1})}+\|\Delta(\phi_{\alpha_n}-\phi^*_{\alpha_n})\|_{L^2(P_{i-1}\cup P_i\cup P_{i+1})})\\ &\le C(\alpha_n-1)\max_{j=1,\dots,l_n}\|\nabla\phi_{\alpha_n}\|_{L^2(P_j)}\\ &\quad+C\big(\|u_n\|_{L^2(P_1)}+\|u_n\|_{L^2(P_{k_n})}\big)\big(e^{-\frac12(k_n-i)L}+e^{-\frac12 iL}\big)\\ &\le C\left((\alpha_n-1)+e^{-\frac12(k_n-i)L}+e^{-\frac12 iL}\right)o(n),\end{aligned}\tag{2.35}$$

where we used the fact that

$$\|u_n\|_{L^2(P_1)}+\|u_n\|_{L^2(P_{k_n})}=o(n).$$

□

With the help of Lemma 2.11 and the Pohozaev type constant in Proposition 2.4, we shall obtain an exponential decay of radial energy and then the whole energy along the first type neck domain. We have the following lemma.

Denoting

$$\Xi_n:=\frac{1}{2\alpha_n-1}\left(4\alpha_n\kappa_n(\phi_{\alpha_n})+\frac{l_n^2}{\pi}\right),$$

we have the following lemma.

LEMMA 2.12. *Under the assumption of Lemma 2.11, let $\lambda \in (0, \frac{1}{4})$ and*

$$\lim_{n\to\infty} \frac{\alpha_n - 1}{l_n} = 0,$$

then for any $t_n \in [T_n^1 + \lambda(T_n^2 - T_n^1), T_n^2 - \lambda(T_n^2 - T_n^1)]$ and $k > 0$, there hold

$$\lim_{n\to\infty} \frac{1}{(\alpha_n - 1) + \frac{1}{(T_n^2 - T_n^1)^k}} \left(\int_{[t_n - k, t_n + k]\times S^1} |\frac{\partial \phi_{\alpha_n}}{\partial \theta}|^2 dt d\theta \right)^{\frac{1}{2}} = 0$$

and

$$\begin{aligned}&\int_{[t_n - k, t_n + k]\times S^1} (1 + |\nabla_{g_n}\phi_{\alpha_n}|^2)^{\alpha_n - 1} |\nabla \phi_{\alpha_n}|^2 dt d\theta \\ &\quad = k\Xi_n + \left(l_n^2 + \frac{1}{(T_n^2 - T_n^1)^2} \right) o(n).\end{aligned}$$

PROOF. By Lemma 2.11, we have

$$\begin{aligned}&\left(\int_{[t_n - k, t_n + k]\times S^1} |\frac{\partial \phi_{\alpha_n}}{\partial \theta}|^2 dt d\theta \right)^{\frac{1}{2}} \\ &\leq C(k) \left((\alpha_n - 1) + e^{-\frac{t_n - k - T_n^1}{2}} + e^{-\frac{T_n^2 - (t_n + k)}{2}} \right) o(n) \\ &\leq C(k) \left((\alpha_n - 1) + e^{-\frac{\lambda(T_n^2 - T_n^1)}{2}} \right) o(n). \qquad (2.36)\end{aligned}$$

Then, we get

$$\begin{aligned}&\frac{1}{(\alpha_n - 1) + \frac{1}{(T_n^2 - T_n^1)^k}} \left(\int_{[t_n - k, t_n + k]\times S^1} |\frac{\partial \phi_{\alpha_n}}{\partial \theta}|^2 dt d\theta \right)^{\frac{1}{2}} \\ &\leq C(k) \left(1 + e^{-\frac{\lambda(T_n^2 - T_n^1)}{2}} (T_n^2 - T_n^1)^k \right) o(n)\end{aligned}$$

and the first conclusion of the lemma follows immediately.

For the second conclusion, by Proposition 2.4, we have

$$\begin{aligned}&\int_{[t_n - k, t_n + k]\times S^1} (1 + |\nabla_{g_n}\phi_{\alpha_n}|^2)^{\alpha_n - 1} |\nabla \phi_{\alpha_n}|^2 dt d\theta \\ &\quad - \frac{2\alpha_n}{2\alpha_n - 1} \int_{[t_n - k, t_n + k]\times S^1} (1 + |\nabla_{g_n}\phi_{\alpha_n}|^2)^{\alpha_n - 1} |\frac{\partial \phi_{\alpha_n}}{\partial \theta}|^2 dt d\theta \\ &= \frac{1}{2\alpha_n - 1} \int_{[t_n - k, t_n + k]\times S^1} (1 + |\nabla_{g_n}\phi_{\alpha_n}|^2)^{\alpha_n - 1} \rho_n^2 dt d\theta \\ &\quad - \frac{2}{2\alpha_n - 1} \int_{t_n - k}^{t_n + k} \int_{[0,s]\times S^1} (1 + |\nabla_{g_n}\phi_{\alpha_n}|^2)^{\alpha_n - 1} \rho_n^3 \sin(\frac{l_n t}{2\pi}) dt d\theta ds \\ &\quad + \frac{2(\alpha_n - 1)}{2\alpha_n - 1} \int_{t_n - k}^{t_n + k} \int_{[0,s]\times S^1} (1 + |\nabla_{g_n}\phi_{\alpha_n}|^2)^{\alpha_n - 1} |\nabla \phi_{\alpha_n}|^2 \rho_n \sin(\frac{l_n t}{2\pi}) dt d\theta ds \\ &\quad + 2k\kappa_n(\phi_{\alpha_n}) \frac{2\alpha_n}{2\alpha_n - 1} \\ &= \mathbb{I} + \mathbb{II} + \mathbb{III} + 2k\kappa_n(\phi_{\alpha_n}) \frac{2\alpha_n}{2\alpha_n - 1}.\end{aligned}$$

To estimate these terms, we need to consider the following cases: $t_n + k \le 0$, $t_n - k \ge 0$ and $t_n - k \le 0 \le t_n + k$. Next, we give the estimates for the case $t_n - k \ge 0$ and another two cases are similar.

CASE 1. $t_n - k \ge 0$.

For the first term $\mathbb{I}$, we have

$$\begin{aligned} \mathbb{I} &= \frac{1}{2\alpha_n - 1} \int_{[t_n-k,t_n+k]\times S^1} \rho_n^2 dt d\theta \\ &\quad + \frac{1}{2\alpha_n - 1} \int_{[t_n-k,t_n+k]\times S^1} \left((1+|\nabla_{g_n}\phi_{\alpha_n}|^2)^{\alpha_n-1} - 1\right) \rho_n^2 dt d\theta \\ &= \mathbb{I}_1 + \mathbb{I}_2. \end{aligned} \tag{2.37}$$

Noting that

$$(1+|\nabla_{g_n}\phi_{\alpha_n}|^2)^{\alpha_n-1} - 1 = o(n),$$

since $\alpha_n - 1 \le C l_n$, by (2.25), we have

$$\begin{aligned} \mathbb{I}_2 &= o(n) \int_{[t_n-k,t_n+k]\times S^1} \rho_n^2 dt d\theta \\ &= l_n \tan(\frac{l_n t}{2\pi})\Big|_{t_n-k}^{t_n+k} o(n) \\ &= l_n \tan(\frac{l_n}{2\pi}2k)\left(1 + \tan\frac{l_n(t_n+k)}{2\pi}\tan\frac{l_n(t_n-k)}{2\pi}\right) o(n) \\ &\le l_n \tan(\frac{l_n}{2\pi}2k)\left(1 + \tan^2\frac{l_n(t_n+k)}{2\pi}\right) o(n). \end{aligned} \tag{2.38}$$

If $\lim_{n\to\infty} \frac{T_n^2 - T_n^1}{T_n} = 0$, noting that

$$\frac{l_n}{2\pi}(t_n+k) \le \frac{l_n}{2\pi}\left(T_n^2 - \lambda(T_n^2 - T_n^1) + k\right) \le \frac{l_n}{2\pi}\left(T_n - \lambda(T_n^2 - T_n^1) + k\right)$$

and

$$\tan(\frac{l_n(t_n+k)}{2\pi}) \le \frac{C}{\lambda l_n(T_n^2 - T_n^1)} \tag{2.39}$$

when n is big enough, then

$$\mathbb{I}_2 = \frac{o(n)}{(T_n^2 - T_n^1)^2}.$$

If $\lim_{n\to\infty} \frac{T_n^2 - T_n^1}{T_n} = C_1 > 0$, noting that $C_1 \le 2$ and

$$\begin{aligned} \frac{l_n}{2\pi}(t_n+k) &\le \frac{l_n}{2\pi}\left(T_n^2 - \lambda(T_n^2 - T_n^1) + k\right) \\ &\le \frac{l_n}{2\pi}\left((1-\lambda C_1)T_n + k\right) \\ &\le (1-\lambda C_1)\frac{\pi}{2} - (1-\lambda C_1)\arctan\sinh(\frac{l_n}{2}) + \frac{l_n k}{2\pi} \end{aligned}$$

and

$$\tan(\frac{l_n(t_n+k)}{2\pi}) \le \tan\left((1-\lambda C_1)\frac{\pi}{2} + \frac{1}{2}\lambda C_1\frac{\pi}{2}\right) \le C \tag{2.40}$$

when n is big enough, then

$$\mathbb{I}_2 = l_n^2 o(n). \tag{2.41}$$

For the second term, we have

$$\begin{aligned}\mathbb{III} = & -\frac{2}{2\alpha_n - 1}\int_{t_n-k}^{t_n+k}\int_{[0,s]\times S^1}\rho_n^3\sin(\frac{l_n t}{2\pi})dtd\theta ds\\ & -\frac{2}{2\alpha_n - 1}\int_{t_n-k}^{t_n+k}\int_{[0,s]\times S^1}\left((1+|\nabla_{g_n}\phi_{\alpha_n}|^2)^{\alpha_n-1}-1\right)\rho_n^3\sin(\frac{l_n t}{2\pi})dtd\theta ds\\ := & \mathbb{III}_1 + \mathbb{III}_2.\end{aligned} \tag{2.42}$$

Similar to estimating the term $\mathbb{I}_2$, we can get

$$\begin{aligned}\mathbb{III}_2 & = -\frac{2}{2\alpha_n - 1}\int_{t_n-k}^{t_n+k}\int_{[0,s]\times S^1}\left((1+|\nabla_{g_n}\phi_{\alpha_n}|^2)^{\alpha_n-1}-1\right)\rho_n^3\sin(\frac{l_n t}{2\pi})dtd\theta ds\\ & = o(n)\int_{t_n-k}^{t_n+k}\int_{[0,s]\times S^1}\rho_n^3\sin(\frac{l_n t}{2\pi})dtd\theta ds\\ & = o(n)\int_{t_n-k}^{t_n+k} l_n^2\tan^2(\frac{l_n t}{2\pi})dt\\ & \le o(n)\int_{t_n-k}^{t_n+k}\rho_n^2 dt = \left(l_n^2 + \frac{1}{(T_n^2 - T_n^1)^2}\right)o(n),\end{aligned} \tag{2.43}$$

where the third equality follows from the fact that

$$\int_0^s \rho_n^3\sin(\frac{l_n t}{2\pi})dt = \frac{1}{2}\rho_n^2\sin^2(\frac{l_n s}{2\pi}), \tag{2.44}$$

Combining (2.37), (2.42) with (2.44), it yields

$$\begin{aligned}\mathbb{I}+\mathbb{III} = & \frac{1}{2\alpha_n - 1}\int_{[t_n-k,t_n+k]\times S^1}\rho_n^2 dtd\theta\\ & -\frac{2}{2\alpha_n - 1}\int_{t_n-k}^{t_n+k}\int_{[0,s]\times S^1}\rho_n^3\sin(\frac{l_n t}{2\pi})dtd\theta ds\\ & +\left(l_n^2 + \frac{1}{(T_n^2 - T_n^1)^2}\right)o(n)\\ = & \frac{1}{2\alpha_n - 1}\int_{[t_n-k,t_n+k]\times S^1}\rho_n^2 dtd\theta - \frac{1}{2\alpha_n - 1}\int_{t_n-k}^{t_n+k}\int_{[0,s]\times S^1}\rho_n^2\sin^2(\frac{l_n t}{2\pi})dtd\theta ds\\ & +\left(l_n^2 + \frac{1}{(T_n^2 - T_n^1)^2}\right)o(n)\\ = & \frac{1}{2\alpha_n - 1}\int_{[t_n-k,t_n+k]\times S^1}\rho_n^2\cos^2(\frac{l_n s}{2\pi})dtd\theta + \left(l_n^2 + \frac{1}{(T_n^2 - T_n^1)^2}\right)o(n)\\ = & \frac{1}{2\alpha_n - 1}\frac{l_n^2}{\pi}k + \left(l_n^2 + \frac{1}{(T_n^2 - T_n^1)^2}\right)o(n).\end{aligned} \tag{2.45}$$

For the third term, since $E_{\alpha_n}(\phi_{\alpha_n}) \le \Lambda$, we get

$$
\begin{aligned}
\frac{|\mathbb{III}|}{\alpha_n - 1} &\le C \int_{t_n-k}^{t_n+k} \int_{[0,s]\times S^1} (1+|\nabla_{g_n}\phi_{\alpha_n}|^2)^{\alpha_n-1}|\nabla\phi_{\alpha_n}|^2 \rho_n \sin(\frac{l_n t}{2\pi}) dt d\theta ds \\
&\le C \int_{t_n-k}^{t_n+k} \int_{[0,s]\times S^1} (1+|\nabla_{g_n}\phi_{\alpha_n}|^2)^{\alpha_n-1}|\nabla\phi_{\alpha_n}|^2 dt d\theta \cdot \rho_n(s) \sin(\frac{l_n s}{2\pi}) ds \\
&\le C \int_{[T_n^1,T_n^2]\times S^1} (1+|\nabla_{g_n}\phi_{\alpha_n}|^2)^{\alpha_n-1}|\nabla\phi_{\alpha_n}|^2 dt d\theta \cdot \int_{t_n-k}^{t_n+k} \rho_n(s) \sin(\frac{l_n s}{2\pi}) ds \\
&\le C\Lambda \log \frac{\cos(\frac{l_n}{2\pi}(t_n-k))}{\cos(\frac{l_n}{2\pi}(t_n+k))} = C\Lambda \log\left(1 + \frac{\cos(\frac{l_n}{2\pi}(t_n-k))}{\cos(\frac{l_n}{2\pi}(t_n+k))} - 1\right) \\
&\le C\Lambda \left(\frac{\cos(\frac{l_n}{2\pi}(t_n-k))}{\cos(\frac{l_n}{2\pi}(t_n+k))} - 1\right) = C\Lambda \left(\frac{\cos(\frac{l_n}{2\pi}(t_n+k-2k))}{\cos(\frac{l_n}{2\pi}(t_n+k))} - 1\right) \\
&= C\Lambda \left(\cos(\frac{l_n}{2\pi}2k) + \tan(\frac{l_n}{2\pi}(t_n+k))\sin(\frac{l_n}{2\pi}2k) - 1\right) \\
&\le C\Lambda \left(|\cos(\frac{l_n}{2\pi}2k) - 1| + |\tan(\frac{l_n}{2\pi}(t_n+k))\sin(\frac{l_n}{2\pi}2k)|\right) \\
&\le C\Lambda \left(l_n^2 k^2 + |\tan(\frac{l_n}{2\pi}(t_n+k))\sin(\frac{l_n}{2\pi}2k)|\right),
\end{aligned}
\tag{2.46}
$$

where we used the fact that

$$\log(1+x) \le x, \ x > 0.$$

Noting that if $\lim_{n\to\infty} \frac{T_n^2 - T_n^1}{T_n} = 0$, by (2.39), there holds

$$\tan\left(\frac{l_n}{2\pi}(t_n+k)\right)\sin(\frac{l_n}{2\pi}2k) \le C(k,\lambda)\frac{1}{T_n^2 - T_n^1};$$

If $\lim_{n\to\infty} \frac{T_n^2 - T_n^1}{T_n} = C_1 > 0$, by (2.40), there holds

$$\tan\left(\frac{l_n}{2\pi}(t_n+k)\right)\sin(\frac{l_n}{2\pi}2k) \le C(k) l_n;$$

Thus, in both cases, we obtain

$$l_n^2 k^2 + \left|\tan\left(\frac{l_n}{2\pi}(t_n+k)\right)\sin(\frac{l_n}{2\pi}2k)\right| \le C\left(l_n + \frac{1}{T_n^2 - T_n^1}\right) \le \frac{C}{T_n^2 - T_n^1},$$

which yields

$$|\mathbb{III}| \le C(k,\lambda)(\alpha_n - 1)\frac{1}{T_n^2 - T_n^1} = C(k,\lambda)\frac{l_n}{T_n^2 - T_n^1} o(n),$$

since $\alpha_n - 1 = o(n) l_n$.

Combining these together, we arrived at

$$\begin{aligned}
&\int_{[t_n-k,t_n+k]\times S^1}(1+|\nabla_{g_n}\phi_{\alpha_n}|^2)^{\alpha_n-1}|\nabla\phi_{\alpha_n}|^2dtd\theta\\
&=\frac{2\alpha_n}{2\alpha_n-1}\int_{[t_n-k,t_n+k]\times S^1}(1+|\nabla_{g_n}\phi_{\alpha_n}|^2)^{\alpha_n-1}|\frac{\partial\phi_{\alpha_n}}{\partial\theta}|^2dtd\theta\\
&+2k\kappa_n(\phi_{\alpha_n})\frac{2\alpha_n}{2\alpha_n-1}+\frac{1}{2\alpha_n-1}\frac{l_n^2}{\pi}k\\
&+\left(l_n^2+\frac{1}{(T_n^2-T_n^1)^2}\right)o(n)\\
&=\frac{1}{2\alpha_n-1}\left(4\alpha_n\kappa_n(\phi_{\alpha_n})+\frac{l_n^2}{\pi}\right)k+\left(l_n^2+\frac{1}{(T_n^2-T_n^1)^2}\right)o(n).
\end{aligned}$$

CASE 2. $t_n+k\le 0$.

This case is similar to **Case 1**.

CASE 3. $t_n-k\le 0\le t_n+k$.

In this case, it is easy to see that $-k\le t_n\le k$. Then by (2.38) and (2.43), we get

$$\mathbb{I}_2+\mathbb{III}_2=o(n)\int_{t_n-k}^{t_n+k}\rho_n^2dt=l_n\tan(\frac{l_n}{2\pi}2k)\left(1+\tan^2\frac{l_n(t_n+k)}{2\pi}\right)o(n)=(l_n)^2o(n);$$

by (2.46), we have

$$\begin{aligned}
\frac{\mathbb{IIII}}{\alpha_n-1}&\le C\int_{[T_n^1,T_n^2]\times S^1}(1+|\nabla_{g_n}\phi_{\alpha_n}|^2)^{\alpha_n-1}|\nabla\phi_{\alpha_n}|^2dtd\theta\\
&\quad\cdot\left(-\int_{t_n-k}^{0}\rho_n(s)\sin(\frac{l_ns}{2\pi})ds+\int_0^{t_n+k}\rho_n(s)\sin(\frac{l_ns}{2\pi})ds\right)\\
&\le -C\log\left(\cos(\frac{l_n}{2\pi}(t_n+k))\cos(\frac{l_n}{2\pi}(t_n-k))\right)\le C(l_n)^2.
\end{aligned}$$

The left proof is the same to **Case 1**. The proof of this lemma is finished. □

From the proof of Lemma 2.12, we can also derive a decay estimate of the energy near the two ends of Q_n, which is useful and necessary to derive the length formula of the limiting neck geodesic.

LEMMA 2.13. *Under the assumption of Lemma 2.11, let*

$$\lim_{n\to\infty}\frac{(1-\alpha_n)\log l_n}{(l_n)^2}<\infty.$$

Then, for any $t_n\in[T_n^1+2,T_n^2-2]$, there holds

$$\begin{aligned}
&\int_{[t_n-2,t_n+2]\times S^1}(1+|\nabla_{g_n}\phi_{\alpha_n}|^2)^{\alpha_n-1}|\nabla\phi_{\alpha_n}|^2dtd\theta\\
&\le 2|\Xi_n|+C(l_n)^2+\left(e^{-(t_n-T_n^1)}+e^{-(T_n^2-t_n)}\right)o(n).
\end{aligned}$$

PROOF. From the proof of Lemma 2.12, using the notations as in Lemma 2.12, we have

$$\begin{aligned}&\int_{[t_n-2,t_n+2]\times S^1}(1+|\nabla_{g_n}\phi_{\alpha_n}|^2)^{\alpha_n-1}|\nabla\phi_{\alpha_n}|^2dtd\theta\\&\quad-\frac{2\alpha_n}{2\alpha_n-1}\int_{[t_n-2,t_n+2]\times S^1}(1+|\nabla_{g_n}\phi_{\alpha_n}|^2)^{\alpha_n-1}|\frac{\partial\phi_{\alpha_n}}{\partial\theta}|^2dtd\theta\\&=\mathbb{I}+\mathbb{II}+\mathbb{IIII}+4\kappa_n(\phi_{\alpha_n})\frac{2\alpha_n}{2\alpha_n-1}=\mathbb{I}_2+\mathbb{II}_2+\mathbb{IIII}+2\Xi_n.\end{aligned}$$

Noting that

$$\sup_{[T_n^1,T_n^2]\times S^1}((1+|\nabla_{g_n}\phi_{\alpha_n}|^2)^{\alpha_n-1}-1)\le C(\alpha_n-1)\log(l_n)^{-2},$$

where we used the fact that

$$x\le 2\log(1+x)$$

when $x>0$ is small enough, by (2.37), (2.42), we get

$$\mathbb{I}_2+\mathbb{II}_2\le C(1-\alpha_n)\log l_n\int_{Q_n}\rho_n^2dtd\theta\le C(1-\alpha_n)\log l_n\le C(l_n)^2.$$

By (2.46), we get that if $t_n-2\ge 0$ or $t_n+2\le 0$,

$$\frac{|\mathbb{IIII}|}{\alpha_n-1}\le C\left|\int_{t_n-2}^{t_n+2}\rho_n\sin\frac{l_nt}{2\pi}dt\right|\le C,$$

where we used the fact that

$$\tan(\frac{l_n}{2\pi}(t_n+2))\sin(\frac{l_n}{2\pi}4)\le\tan(\frac{l_n}{2\pi}T_n)\sin(\frac{l_n}{2\pi}4)\le C;$$

if $t_n-2\le 0\le t_n+2$,

$$\begin{aligned}\frac{|\mathbb{IIII}|}{\alpha_n-1}&\le C\left(-\int_{t_n-2}^{0}\rho_n\sin\frac{l_nt}{2\pi}dt+\int_0^{t_n+2}\rho_n\sin\frac{l_nt}{2\pi}dt\right)\\&\le C\left(-\int_{-4}^{0}\rho_n\sin\frac{l_nt}{2\pi}dt+\int_0^{4}\rho_n\sin\frac{l_nt}{2\pi}dt\right)\le C;\end{aligned}$$

which imply that

$$|\mathbb{IIII}|\le C(\alpha_n-1)\le C\frac{(l_n)^2}{|\log l_n|}\le C(l_n)^2.$$

Combining this with (2.36), we get the conclusion of the lemma. □

From Lemma 2.12 and Lemma 2.13, we immediately have the following no neck theorem.

THEOREM 2.14. *Under the assumptions of Lemma 2.13, if*

$$\lim_{n\to\infty}\sqrt{|\Xi_n|}(T_n^2-T_n^1)=0,$$

then

$$\lim_{n\to\infty}Osc_{[T_n^1,T_n^2]\times S^1}\phi_{\alpha_n}=0.$$

PROOF. By Lemma 2.12, we know that for any $t_n \in [T_n^1 + \lambda(T_n^2 - T_n^1), T_n^2 - \lambda(T_n^2 - T_n^1)]$, there holds

$$\int_{[t_n-2,t_n+2]\times S^1} |\nabla\phi_{\alpha_n}|^2 dtd\theta \le \int_{[t_n-2,t_n+2]\times S^1} (1+|\nabla_{g_n}\phi_{\alpha_n}|^2)^{\alpha_n-1}|\nabla\phi_{\alpha_n}|^2 dtd\theta$$
$$\le 2|\Xi_n| + \frac{o(n)}{(T_n^2 - T_n^1)^2}.$$

Then, Lemma 2.5 tells us that

$$Osc_{[t_n-1,t_n+1]\times S^1}\ \phi_{\alpha_n} \le C\|\nabla\phi_{\alpha_n}\|_{L^2([t_n-2,t_n+2]\times S^1)} \le \sqrt{2\Xi_n} + \frac{o(n)}{T_n^2 - T_n^1},$$

which implies

$$Osc_{[T_n^1+\lambda(T_n^2-T_n^1),T_n^2-\lambda(T_n^2-T_n^1)]\times S^1}\ \phi_{\alpha_n} \le \left(\sqrt{2|\Xi_n|} + \frac{o(n)}{T_n^2 - T_n^1}\right)\frac{(1-2\lambda)(T_n^2 - T_n^1)}{2}.$$

To estimate the oscillation on

$$[T_n^1 + 1, T_n^1 + \lambda(T_n^2 - T_n^1)] \cup [T_n^2 - \lambda(T_n^2 - T_n^1), T_n^2 - 1],$$

recalling that $P_i = [T_n^1 + (i-1)L, T_n^1 + iL] \times S^1$, by Lemma 2.13, we have

$$\int_{P_i} (1+|\nabla_{g_n}\phi_{\alpha_n}|^2)^{\alpha_n-1}|\nabla\phi_{\alpha_n}|^2 dtd\theta \le 2|\Xi_n| + C(l_n)^2 + e^{-iL}o(n)$$

if $2 \le i \le [\frac{\lambda(T_n^2-T_n^1)}{L}] + 1$ and

$$\int_{P_i} (1+|\nabla_{g_n}\phi_{\alpha_n}|^2)^{\alpha_n-1}|\nabla\phi_{\alpha_n}|^2 dtd\theta \le 2|\Xi_n| + C(l_n)^2 + e^{T_n^2-T_n^1-iL}o(n)$$

if $[\frac{(1-\lambda)(T_n^2-T_n^1)}{L}] - 1 \le i \le [\frac{T_n^2-T_n^1-1}{L}]$, where [] means that we take the integer part of a positive constant.

Then, Lemma 2.5 tells us that for $2 \le i \le [\frac{\lambda(T_n^2-T_n^1)}{L}] + 1$, there holds

$$Osc_{P_i}\phi_{\alpha_n} \le C\|\nabla\phi_{\alpha_n}\|_{L^2(P_{i-1}\cup P_i\cup P_{i+1})} \le 3\sqrt{2\Xi_n} + Cl_n + e^{-\frac{iL}{2}}o(n),$$

which implies

$$Osc_{[T_n^1+1,T_n^1+\lambda(T_n^2-T_n^1)]\times S^1}\phi_{\alpha_n} \le \sum_{i=2}^{[\frac{\lambda(T_n^2-T_n^1)}{L}]+1} Osc_{P_i}\phi_{\alpha_n}$$
$$\le C\left(\sqrt{|\Xi_n|} + l_n\right)\frac{\lambda(T_n^2 - T_n^1)}{L} + o(n).$$

Similarly,

$$Osc_{[T_n^2-\lambda(T_n^2-T_n^1),T_n^2-1]\times S^1}\phi_{\alpha_n} \le C\left(\sqrt{|\Xi_n|} + l_n\right)\frac{\lambda(T_n^2 - T_n^1)}{L} + o(n).$$

By letting firstly $n \to \infty$, then $\lambda \to 0$, we get the conclusion of this theorem immediately. □

Generally, the quantity

$$\lim_{n\to\infty} \sqrt{|\Xi_n|}(T_n^2 - T_n^1)$$

may not be zero, then we want to study the limit of the neck domain. For this, we first show the following lemma.

LEMMA 2.15. *Under the assumption of Lemma 2.11, let* $\lambda \in (0, \frac{1}{4})$ *and*

$$\lim_{n\to\infty} \sqrt{|\Xi_n|}(T_n^2 - T_n^1) > 0, \quad \lim_{n\to\infty} \frac{\alpha_n - 1}{l_n} = 0, \tag{2.47}$$

then there holds

$$\lim_{n\to\infty} \left\| \frac{\phi_{\alpha_n} - \phi_{\alpha_n}^*}{\sqrt{|\Xi_n|}} \right\|_{C^1([T_n^1+\lambda(T_n^2-T_n^1), T_n^2-\lambda(T_n^2-T_n^1)]\times S^1)} = 0,$$

where

$$\phi_{\alpha_n}^*(t) := \frac{1}{2\pi}\int_0^{2\pi} \phi_{\alpha_n}(t,\theta)d\theta.$$

PROOF. If it was false, then there would exist $t_n \in [T_n^1 + \lambda(T_n^2 - T_n^1), T_n^2 - \lambda(T_n^2 - T_n^1)]$ and $\theta_n \in [0, 2\pi]$ such that

$$\frac{1}{\sqrt{|\Xi_n|}} \left(|\nabla\phi_{\alpha_n}(t_n, \theta_n) - \nabla\phi_{\alpha_n}^*(t_n)| + |\phi_{\alpha_n}(t_n,\theta_n) - \phi_{\alpha_n}^*(t_n)| \right) \to b \neq 0. \tag{2.48}$$

By (2.47) and Lemma 2.12, we have

$$\frac{1}{T_n^2 - T_n^1} \leq C\sqrt{|\Xi_n|}, \ and \ \frac{1}{\sqrt{|\Xi_n|}} \leq C(T_n^2 - T_n^1) \tag{2.49}$$

and for any $k > 0$, there holds

$$\int_{[t_n-k, t_n+k]\times S^1} |\nabla\phi_{\alpha_n}|^2 dtd\theta \leq C(k,\lambda)|\Xi_n|,$$

when n is big enough.

Then Lemma 2.5 tells us that

$$\|\nabla\phi_{\alpha_n}\|_{C^1([t_n-k,t_n+k]\times S^1)} \leq C\sqrt{|\Xi_n|}. \tag{2.50}$$

Setting

$$u_n(t,\theta) := \frac{1}{\sqrt{|\Xi_n|}} \left(\phi_{\alpha_n}(t_n + t, \theta) - \phi_{\alpha_n}(t_n, 0)\right),$$

then, it is easy to see that

$$\|\nabla u_n\|_{C^1([-k,k]\times S^1)} \leq C(k). \tag{2.51}$$

Combining this with the fact that $u_n(0,0) = 0$, we have

$$\|u_n\|_{C^2([-k,k]\times S^1)} \leq C(k).$$

From (2.18), (2.19) and (2.20), it is easy to see that u_n satisfies the following equation

$$\Delta u_n + \sqrt{|\Xi_n|}O(|\nabla u_n|^2) + (\alpha_n - 1)O(|\nabla^2 u_n|) + (\alpha_n - 1)O(|\nabla u_n|) = 0.$$

Therefore, we know that after passing to a subsequence, there holds

$$u_n \to u \in C^1_{loc}(\mathbb{R}^1 \times S^1)$$

where u is a harmonic function, i.e. $\Delta u = 0$.

Moreover, by Lemma 2.12, we get

$$\int_{[-k,k]\times S^1} |\frac{\partial u}{\partial\theta}|^2 dtd\theta = \lim_{n\to\infty} \frac{1}{|\Xi_n|}\int_{[t_n-k,t_n+k]\times S^1} |\frac{\partial\phi_{\alpha_n}}{\partial\theta}|^2 dtd\theta = 0,$$

where we used the fact that

$$\frac{\alpha_n - 1 + \frac{1}{(T_n^2-T_n^1)^k}}{|\Xi_n|} \le \frac{l_n + \frac{1}{(T_n^2-T_n^1)^k}}{|\Xi_n|} \le \frac{1}{|\Xi_n|(T_n^2 - T_n^1)} \le C.$$

This implies

$$\frac{\partial u}{\partial \theta} = 0.$$

Thus, we know that the harmonic function u must be of the form

$$u(t) = \overrightarrow{a} t,$$

where $\overrightarrow{a} = (a^1, \dots, a^K) \in T_y N$. Let $\lim_{n\to\infty} \theta_n = \theta_0$, then it is obvious that

$$|\nabla u(0,\theta_0) - \nabla u^*(0)| + |u(0,\theta_0) - u^*(0)| = 0.$$

However, by (2.48), we have

$$\begin{aligned} b &= \lim_{n\to\infty} \frac{1}{\sqrt{|\Xi_n|}} \left(|\nabla \phi_n(t_n,\theta_n) - \nabla \phi_n^*(t_n)| + |\phi_n(t_n,\theta_n) - \phi_n^*(t_n)| \right) \\ &= \lim_{n\to\infty} |\nabla u_n(0,\theta_n) - \nabla u_n^*(0)| + \lim_{n\to\infty} |u_n(0,\theta_n) - u_n^*(0)| \\ &= |\nabla u(0,\theta_0) - \nabla u^*(0)| + |u(0,\theta_0) - u^*(0)| = 0, \end{aligned}$$

which is a contradiction. So, we finished the proof of the lemma. □

LEMMA 2.16. *Under the assumption of Lemma 2.15, there holds*

$$\left\| \frac{1}{\sqrt{|\Xi_n|}} |\nabla \phi_{\alpha_n}| - \frac{1}{\sqrt{4\pi}} \right\|_{C^0([T_n^1+\lambda(T_n^2-T_n^1), T_n^2-\lambda(T_n^2-T_n^1)]\times S^1)} \to 0. \tag{2.52}$$

PROOF. If the conclusion does not hold, then there exist $t_n \in [T_n^1 + \lambda(T_n^2 - T_n^1), T_n^2 - \lambda(T_n^2 - T_n^1)]$ and $\theta_n \in [0, 2\pi]$ such that

$$\frac{1}{\sqrt{|\Xi_n|}} |\nabla \phi_{\alpha_n}(t_n,\theta_n)| \to d \neq \frac{1}{\sqrt{4\pi}}.$$

By the argument in the proof of Lemma 2.15, we may assume

$$u_n := \frac{1}{\sqrt{|\Xi_n|}} \left(\phi_{\alpha_n}(t_n + t,\theta) - \phi_{\alpha_n}(t_n, 0) \right) \to u(t) \textit{ in } C^1_{loc}(\mathbb{R}^1 \times S^1),$$

where

$$u(t) = \overrightarrow{a} t$$

with $|\overrightarrow{a}| = d$. Then,

$$\lim_{n\to\infty} \int_{[-1,1]\times S^1} |\nabla u_n|^2 dt d\theta = 4\pi |\overrightarrow{a}|^2. \tag{2.53}$$

On the other hand, by Lemma 2.12, we get

$$\begin{aligned}
&\lim_{n\to\infty}\int_{[-1,1]\times S^1}|\nabla u_n|^2dtd\theta\\
&=\lim_{n\to\infty}\frac{1}{|\Xi_n|}\int_{[t_n-1,t_n+1]\times S^1}|\nabla\phi_{\alpha_n}|^2dtd\theta\\
&=\lim_{n\to\infty}\frac{1}{|\Xi_n|}\int_{[t_n-1,t_n+1]\times S^1}(1+|\nabla_{g_n}\phi_{\alpha_n}|^2)^{\alpha_n-1}|\nabla\phi_{\alpha_n}|^2dtd\theta\\
&=1+\lim_{n\to\infty}\frac{1}{|\Xi_n|}\frac{o(n)}{(T_n^2-T_n^1)^2}=1,
\end{aligned}$$

where the last inequality follows from (2.49).

Therefore, we get

$$d=|\overrightarrow{d}|=\frac{1}{\sqrt{4\pi}},$$

which is also a contradiction. We finished the proof of this lemma. □

Now, we give the proof of Theorem 2.2.

Proof of Theorem 2.2. The conclusion (1) follows directly from Theorem 2.14. Next, we prove the other two conclusions. Recall the notation

$$\Xi:=\lim_{n\to\infty}\sqrt{|\Xi_n|}(T_n^2-T_n^1)$$

and we divide the proof into two steps.

STEP 1. $\Xi\in(0,\infty)$.

Denote

$$\gamma_n:\quad \phi_n^*(t)=\frac{1}{2\pi}\int_0^{2\pi}\phi_{\alpha_n}(t,\theta)d\theta,$$

where $t\in[T_n^1+\lambda(T_n^2-T_n^1),T_n^2-\lambda(T_n^2-T_n^1)]$, $\lambda\in(0,\frac{1}{4})$.

By Lemma 2.5, it is easy to see that γ_n converges to some curve on N denoted by γ. Next, we will show that γ is a geodesic on N.

Denoting

$$\dot{\phi}^*_{\alpha_n}=\frac{d\phi^*_{\alpha_n}}{dt},\ \ddot{\phi}^*_{\alpha_n}=\frac{d^2\phi^*_{\alpha_n}}{dt^2},$$

by a direct computation and (2.18), we have

$$\begin{aligned}
\ddot{\phi}^*_{\alpha_n}(t)&=\frac{1}{2\pi}\int_0^{2\pi}\frac{\partial^2\phi_{\alpha_n}}{\partial t^2}(t,\theta)d\theta=\frac{1}{2\pi}\int_0^{2\pi}\Delta\phi_{\alpha_n}d\theta\\
&=-\frac{1}{2\pi}\int_0^{2\pi}A(\phi_n)(\nabla\phi_{\alpha_n},\nabla\phi_{\alpha_n})d\theta\\
&\quad-(\alpha_n-1)\frac{1}{2\pi}\int_0^{2\pi}\frac{\frac{1}{\rho_n^2}\nabla|\nabla\phi_{\alpha_{\alpha_n}}|^2\nabla\phi_{\alpha_{\alpha_n}}}{1+\frac{1}{\rho_n^2}|\nabla\phi_{\alpha_{\alpha_n}}|^2}d\theta\\
&\quad-(\alpha_n-1)\frac{1}{2\pi}\int_0^{2\pi}\frac{\nabla(\frac{1}{\rho_n^2})|\nabla\phi_{\alpha_n}|^2\nabla\phi_{\alpha_n}}{1+\frac{1}{\rho_n^2}|\nabla\phi_{\alpha_n}|^2}d\theta.
\end{aligned}$$

For any $t_n \in [T_n^1+\lambda(T_n^2-T_n^1), T_n^2-\lambda(T_n^2-T_n^1)]$, let s be the arc length parameter of $\phi_n^*(t)$ based on the point t_n, i.e.

$$s(t) = \int_{t_n}^{t} |\dot{\phi}_n^*(\sigma)| d\sigma.$$

Noting that

$$[t_n - \frac{\lambda}{2}(T_n^2 - T_n^1), t_n + \frac{\lambda}{2}(T_n^2 - T_n^1)] \subset [T_n^1 + \frac{\lambda}{2}(T_n^2 - T_n^1), T_n^2 - \frac{\lambda}{2}(T_n^2 - T_n^1)],$$

then by Lemma 2.15 and Lemma 2.16, we have

$$\int_{t_n - \frac{\lambda}{2}(T_n^2 - T_n^1)}^{t_n + \frac{\lambda}{2}(T_n^2 - T_n^1)} |\dot{\phi}_{\alpha_n}^*| dt = \lambda(T_n^2 - T_n^1)\sqrt{|\Xi_n|}\left(\frac{1}{\sqrt{4\pi}} + o(1)\right), \ as\ n \to \infty.$$

Thus, for any $s \in (0, \frac{\lambda}{\sqrt{4\pi}}\Xi)$, there exists $\lambda' \in (0, \lambda)$ such that

$$\phi_{\alpha_n}^*|_{[-s,s]} \subset \phi_{\alpha_n}^*\left([T_n^1 + \frac{\lambda'}{2}(T_n^2 - T_n^1), T_n^2 - \frac{\lambda'}{2}(T_n^2 - T_n^1)]\right),$$

when n is big enough.

Computing directly, we obtain

$$\begin{aligned}
\frac{d^2\phi_{\alpha_n}^*}{ds^2} &= -A_{\gamma_n}(\phi_{\alpha_n}^*(s))(\frac{d\phi_{\alpha_n}^*}{ds}, \frac{d\phi_{\alpha_n}^*}{ds}) = -\frac{1}{|\dot{\phi}_{\alpha_n}^*|^2} A_{\gamma_n}(\phi_{\alpha_n}^*(t))(\frac{d\phi_{\alpha_n}^*}{dt}, \frac{d\phi_{\alpha_n}^*}{dt}) \\
&= \frac{1}{|\dot{\phi}_{\alpha_n}^*|^2}(\ddot{\phi}_{\alpha_n}^* - \frac{\langle \ddot{\phi}_{\alpha_n}^*, \dot{\phi}_{\alpha_n}^* \rangle}{|\dot{\phi}_{\alpha_n}^*|^2}\dot{\phi}_{\alpha_n}^*) \\
&= -\frac{1}{|\dot{\phi}_{\alpha_n}^*|^2}\frac{1}{2\pi}\int_0^{2\pi} A(\phi_{\alpha_n})(\nabla\phi_{\alpha_n}, \nabla\phi_{\alpha_n}) d\theta \\
&\quad + \frac{\dot{\phi}_{\alpha_n}^*}{|\dot{\phi}_{\alpha_n}^*|^4}\frac{1}{2\pi}\int_0^{2\pi} \langle A(\phi_{\alpha_n})(\nabla\phi_{\alpha_n}, \nabla\phi_{\alpha_n}), \dot{\phi}_{\alpha_n}^* \rangle d\theta \\
&\quad - \frac{1}{|\dot{\phi}_{\alpha_n}^*|^2}(\alpha_n - 1) \\
&\quad \times \frac{1}{2\pi}\left(\int_0^{2\pi} \frac{\frac{1}{\rho_n^2}\nabla|\nabla\phi_{\alpha_n}|^2\nabla\phi_{\alpha_n}}{1 + \frac{1}{\rho_n^2}|\nabla\phi_{\alpha_n}|^2} d\theta + \int_0^{2\pi} \frac{\nabla(\frac{1}{\rho_n^2})|\nabla\phi_{\alpha_n}|^2\nabla\phi_{\alpha_n}}{1 + \frac{1}{\rho_n^2}|\nabla\phi_{\alpha_n}|^2} d\theta\right) \\
&\quad + \frac{\dot{\phi}_{\alpha_n}^*}{|\dot{\phi}_{\alpha_n}^*|^4}(\alpha_n - 1) \\
&\quad \times \frac{1}{2\pi}\int_0^{2\pi} \langle \frac{\frac{1}{\rho_n^2}\nabla|\nabla\phi_{\alpha_n}|^2\nabla\phi_{\alpha_n}}{1 + \frac{1}{\rho_n^2}|\nabla\phi_{\alpha_n}|^2} + \frac{\nabla(\frac{1}{\rho_n^2})|\nabla\phi_{\alpha_n}|^2\nabla\phi_{\alpha_n}}{1 + \frac{1}{\rho_n^2}|\nabla\phi_{\alpha_n}|^2}, \dot{\phi}_{\alpha_n}^* \rangle d\theta \\
&:= \mathbf{I} + \mathbf{II} + \mathbf{III} + \mathbf{IV}.
\end{aligned} \tag{2.54}$$

Here, A_{γ_n} is the second fundamental form of $\gamma_n \subset \mathbb{R}^N$.

On the one hand, by Lemma 2.15, Lemma 2.16 and using the fact that

$$\langle A(\phi_{\alpha_n})(\nabla\phi_{\alpha_n}, \nabla\phi_{\alpha_n}), \frac{\partial\phi_{\alpha_n}}{\partial t} \rangle = 0,$$

we have

$$\begin{aligned}\mathbf{II} &= \frac{\dot{\phi}^*_{\alpha_n}}{|\dot{\phi}^*_{\alpha_n}|^4}\frac{1}{2\pi}\int_0^{2\pi}\langle A(\phi_{\alpha_n})(\nabla\phi_{\alpha_n},\nabla\phi_{\alpha_n}),\dot{\phi}^*_{\alpha_n}-\frac{\partial\phi_{\alpha_n}}{\partial t}\rangle d\theta\\ &= \frac{\dot{\phi}^*_{\alpha_n}}{|\dot{\phi}^*_{\alpha_n}|^4}\left(\sqrt{|\Xi_n|}\right)^3 o(1)\to 0 \qquad (2.55)\end{aligned}$$

as $n\to\infty$.

On the other hand, also from Lemma 2.15, Lemma 2.16, (2.19) (2.20), (2.49) and (2.50), it is easy to see that

$$\begin{aligned}\mathbf{III}+\mathbf{IV} &\le C\frac{\alpha_n-1}{|\dot{\phi}^*_{\alpha_n}|^2}\big(\|\nabla^2\phi_{\alpha_n}\|_{L^\infty([T_n^1+\frac{\lambda}{2}(T_n^2-T_n^1),T_n^2-\frac{\lambda}{2}(T_n^2-T_n^1)]\times S^1)}\\ &\qquad + \|\nabla\phi_{\alpha_n}\|_{L^\infty([T_n^1+\frac{\lambda}{2}(T_n^2-T_n^1),T_n^2-\frac{\lambda}{2}(T_n^2-T_n^1)]\times S^1)}\big)\\ &\le C\frac{\alpha_n-1}{\sqrt{|\Xi_n|}}\le C\frac{\alpha_n-1}{l_n}\to 0 \ as\ n\to\infty.\end{aligned}$$

Therefore, we obviously see that

$$|\frac{d^2\phi^*_{\alpha_n}}{ds^2}|\le C(N).$$

Thus, $\phi^*_{\alpha_n}(s)$ will converge locally to a smooth vector valued function from $[-s,s]$ into $\mathbb{R}^K$, denoted by $\omega(s)$, in the sense of C^1, i.e. $\gamma_n|_{[T_n^1+\lambda(T_n^2-T_n^1),T_n^2-\lambda(T_n^2-T_n^1)]}$ converges locally to the curve γ.

Next, we will show that γ is a geodesic.

By Lemma 2.15 and Lemma 2.16, we obtain

$$\begin{aligned}-\mathbf{I} &= \frac{1}{|\dot{\phi}^*_{\alpha_n}|^2}\frac{1}{2\pi}\int_0^{2\pi}A(\phi_{\alpha_n})(\nabla\phi_{\alpha_n},\nabla\phi_{\alpha_n})d\theta\\ &= \frac{1}{|\dot{\phi}^*_{\alpha_n}|^2}\frac{1}{2\pi}\Big(\int_0^{2\pi}(A(\phi_{\alpha_n})\\ &\qquad - A(\phi^*_{\alpha_n}))(\nabla\phi_{\alpha_n},\nabla\phi_{\alpha_n})+A(\phi^*_{\alpha_n})(\nabla(\phi_{\alpha_n}-\phi^*_{\alpha_n}),\nabla\phi_{\alpha_n})d\theta\Big)\\ &\quad +\frac{1}{|\dot{\phi}^*_{\alpha_n}|^2}\frac{1}{2\pi}\int_0^{2\pi}A(\phi^*_{\alpha_n})(\nabla\phi^*_{\alpha_n},\nabla(\phi_{\alpha_n}-\phi^*_{\alpha_n}))d\theta\\ &\quad +\frac{1}{|\dot{\phi}^*_{\alpha_n}|^2}\frac{1}{2\pi}\int_0^{2\pi}A(\phi^*_{\alpha_n})(\nabla\phi^*_{\alpha_n},\nabla\phi^*_{\alpha_n})d\theta\\ &= \frac{1}{|\dot{\phi}^*_{\alpha_n}|^2}A(\phi^*_{\alpha_n})(\nabla\phi^*_{\alpha_n},\nabla\phi^*_{\alpha_n})+\frac{|\Xi_n|}{|\dot{\phi}^*_{\alpha_n}|^2}o(n). \qquad (2.56)\end{aligned}$$

Combining these together, we have

$$\frac{d\omega}{ds}(s)-\frac{d\omega}{ds}(0)=-\int_0^s A(\omega)(\frac{d\omega}{ds},\frac{d\omega}{ds})$$

which implies

$$\frac{d^2\omega}{ds^2}=-A(\omega)(\frac{d\omega}{ds},\frac{d\omega}{ds}).$$

Now, we compute the length of γ. By Lemma 2.15 and Lemma 2.16, we have

$$L(\gamma_n|_{[T_n^1+\lambda(T_n^2-T_n^1),T_n^2-\lambda(T_n^2-T_n^1)]}) = \sqrt{|\Xi_n|}(T_n^2 - T_n^1)(1-2\lambda)\left(\frac{1}{\sqrt{4\pi}} + o(n)\right).$$

Recalling that $P_i = [T_n^1 + (i-1)L, T_n^1 + iL]$ and without loss of generality, we may assume $\lambda(T_n^2 - T_n^1) = k_n L$ for some positive integer k_n, which goes to infinity as $n \to \infty$. Then according to Lemma 2.13 and Lemma 2.5, we obtain

$$\begin{aligned} Osc_{[T_n^1,T_n^1+\lambda(T_n^2-T_n^1)]\times S^1}\phi_{\alpha_n} &\le \sum_{i=1}^{k_n} Osc_{P_i}\phi_{\alpha_n} \\ &\le C\sum_{i=1}^{k_n} \|\nabla\phi_{\alpha_n}\|_{L^2(P_{i-1}\cup P_i\cup P_{i+1})} \\ &\le C\sqrt{|\Xi_n|}(T_n^2 - T_n^1)\lambda + \sum_{i=1}^{k_n}(e^{-\frac{iL}{2}} + e^{\frac{(k_n-i)L}{2}})o(n), \end{aligned}$$

which implies

$$\lim_{\lambda\to 0}\lim_{n\to\infty} Osc_{[T_n^1,T_n^1+\lambda(T_n^2-T_n^1)]\times S^1}\phi_{\alpha_n} = 0.$$

Similarly, there holds

$$\lim_{\lambda\to 0}\lim_{n\to\infty} Osc_{[T_n^2-\lambda(T_n^2-T_n^1),T_n^2]\times S^1}\phi_{\alpha_n} = 0.$$

Therefore, we conclude that

$$\lim_{n\to\infty} L(\gamma_n|_{[T_n^1,T_n^2]}) = \lim_{\lambda\to 0}\lim_{n\to\infty} L(\gamma_n|_{[T_n^1+\lambda(T_n^2-T_n^1),T_n^2-\lambda(T_n^2-T_n^1)]}) = \frac{1}{\sqrt{4\pi}}\Xi.$$

STEP 2. $\Xi = \infty$.

In this case, we prove that the limit contains an infinity length geodesic. In fact, by Lemma 2.15 and Lemma 2.16, we have

$$\begin{aligned} L(\gamma) &\ge \lim_{n\to\infty}\int_{T_n^1+\lambda(T_n^2-T_n^1)}^{T_n^2-\lambda(T_n^2-T_n^1)} |\dot{\phi}^*_{\alpha_n}|dt \\ &= \lim_{n\to\infty}(1-2\lambda)(T_N^2 - T_n^1)\sqrt{|\Xi_n|}\left(\frac{1}{\sqrt{4\pi}} + o(1)\right) = \infty. \end{aligned}$$

We finished the proof of this theorem. □

2.5. Compactness with bounded Morse index

Let $\phi : (\mathbb{R}\times S^1, g) \to (N, h)$ be a α-harmonic map. $\Gamma = \phi^*TN$ is a smooth pull-back bundle over $\mathbb{R}\times S^1$. Let V be a section of Γ and

$$\phi_\tau(x) = exp_{\phi(x)}(\tau V).$$

Obviously, $\phi_0 = \phi$. Then, the formula of the second variation of E_α reads

$$
\begin{aligned}
\delta^2 E_\alpha(\phi)(V,V) &= \frac{d^2}{d\tau^2} E_\alpha(\phi_\tau)|_{\tau=0} \\
&= 2\alpha \int_{\mathbb{R}\times S^1} (1+|\nabla_g\phi|^2)^{(\alpha-1)} \left(\langle \nabla_g V, \nabla_g V\rangle\right. \\
&\quad \left. - R(V, \nabla_g\phi, \nabla_g\phi, V)\right) dvol_g \\
&\quad + 4\alpha(\alpha-1) \int_{\mathbb{R}\times S^1} (1+|\nabla_g\phi|^2)^{\alpha-2} \langle \nabla_g\phi, \nabla_g V\rangle^2 dvol_g. \qquad (2.57)
\end{aligned}
$$

For more details we refer to [**17, 20**].

Let $\mathcal{F}$ denotes the linear space of the smooth sections of Γ. Then, the Morse index of ϕ is the maximal dimension of the linear subspaces of $\mathcal{F}$ on which the (2.57) is negative definite.

PROPOSITION 2.17. *Let (N, g_N) be a compact Riemannian manifold such that one of the following two assumptions holds:*

(i) *the Ricci curvature of the target manifold (N, g_N) has a positive lower bound, i.e. there exists a positive constant $\lambda_N > 0$ such that $Ric_N \geq \lambda_N > 0$;*
(ii) *the target manifold (N, g_N) has a finite fundamental group.*

Let ϕ_{α_n} be a sequence of α_n-harmonic maps from (Q_n, g_n) to N with uniformly bounded α-energy $E_{\alpha_n}(\phi_{\alpha_n}) \leq \Lambda < \infty$, which satisfies

$$\lim_{n\to\infty} \sup_{t\in[T_n^1, T_n^2]} E(\phi_{\alpha_n}, [t-1, t+1]\times S^1) = 0$$

and

$$\lim_{n\to\infty} \frac{\alpha_n - 1}{l_n} = 0, \qquad (2.58)$$

where $-T_n + 2 \leq T_n^1 < T_n^2 \leq T_n - 2$. If

$$\Xi = \infty,$$

then the Morse index of ϕ_{α_n} tends to infinity.

Recalling the curve

$$\phi_{\alpha_n}^*(t) = \frac{1}{2\pi}\int_0^{2\pi} \phi_{\alpha_n}(t,\theta) d\theta,$$

let $s = s(t)$ be the arc-length parametrization of the curve $\phi_{\alpha_n}^*(t)$ with $s(T_n^1 + \lambda(T_n^2 - T_n^1)) = 0$, i.e.

$$s(t) = \int_{T_n^1+\lambda(T_n^2-T_n^1)}^{t} |\dot{\phi}_{\alpha_n}^*|(\sigma) d\sigma. \qquad (2.59)$$

Set $s(t_n^a) = a$, by Theorem 2.2, we know that $\phi_{\alpha_n}^*(s)$ converges to γ on $[0, a]$. Moreover, by Lemma 2.16, we have

$$
\begin{aligned}
a = s(t_n^a) &= \int_{T_n^1+\lambda(T_n^2-T_n^1)}^{t_n^a} |\phi_{\alpha_n}^*|(t) dt \qquad (2.60)\\
&= \left(\frac{1}{\sqrt{4\pi}} + o(1)\right) \sqrt{|\Xi_n|} \left(t_n^a - (T_n^1 + \lambda(T_n^2 - T_n^1))\right)
\end{aligned}
$$

which implies

$$\lim_{n\to\infty} t_n^a - \left(T_n^1 + \lambda(T_n^2 - T_n^1)\right) = \infty \tag{2.61}$$

$$\text{and} \quad \sqrt{|\Xi_n|}\left(t_n^a - \left(T_n^1 + \lambda(T_n^2 - T_n^1)\right)\right) \le C(a).$$

LEMMA 2.18. *If the limiting neck of ϕ_{α_n} is an unstable geodesic which is parameterised on $[0,a]$ by arc length, then for sufficiently large n, there exists a section V_n of $\phi_{\alpha_n}^*(TN)$, which is supported in $[T_n^1 + \lambda(T_n^2 - T_n^1), t_n^a]$, such that*

$$\delta^2 E_{\alpha_n}(\phi_{\alpha_n})(V_n, V_n) < 0.$$

PROOF. Since the limiting neck of $\{\phi_{\alpha_n}\}$, denoted by $\gamma : [0,\, a] \to N$, is not a stable geodesic, there exists a vector field V_0 on γ with $V_0|_{\gamma(0)} = 0$ and $V_0|_{\gamma(a)} = 0$ such that

$$I_\gamma(V_0, V_0) = \int_0^a \left(\langle \nabla_{\dot\gamma} V_0, \nabla_{\dot\gamma} V_0\rangle - R(V_0, \dot\gamma, \dot\gamma, V_0)\right) ds < -\delta < 0. \tag{2.62}$$

Let P be projection from $T\mathbb{R}^K$ to TN. We define

$$V_n(s,\theta) = P_{\phi_{\alpha_n}(s,\theta)}(V_0(s)),$$

where s is the arc-length parametrization of $\phi_{\alpha_n}^*(t)$ given by (2.59). Then, it is easy to see that V_n is a smooth section of $\phi_{\alpha_n}^*(TN)$ which is supported in $[0,\, t_n^a] \times S^1$. Since $\phi_{\alpha_n}(s,\theta) \to \gamma(s)$ in $C^1([0,a] \times S^1)$, we have

$$V_n(\phi_{\alpha_n}(s,\theta)) \to V_0(s) \text{ in } C^1([0,\, a] \times S^1).$$

Next, we will show

$$\lim_{n\to+\infty} \frac{1}{\sqrt{|\Xi_n|}} \delta^2 E_{\alpha_n}(\phi_{\alpha_n})(V_n, V_n) = 2\sqrt{\pi} I_\gamma(V_0, V_0). \tag{2.63}$$

By (2.57), we have

$$\begin{aligned}
\delta^2 E_{\alpha_n}(\phi_{\alpha_n})(V_n, V_n) &= \frac{d^2}{d\tau^2}\Big|_{\tau=0} E_{\alpha_n}((\phi_{\alpha_n})_\tau) \\
&= 2\alpha_n \int_{[T_n^1, T_n^2]\times S^1} (1 + |\nabla_{g_n}\phi_{\alpha_n}|^2)^{(\alpha_n - 1)} \left(\langle \nabla_{g_n} V_n, \nabla_{g_n} V_n\rangle \right. \\
&\quad \left. - R(V_n, \nabla_{g_n}\phi_{\alpha_n}, \nabla_{g_n}\phi_{\alpha_n}, V_n)\right) dvol_{g_n} \\
&\quad + 4\alpha_n(\alpha_n - 1) \int_{[T_n^1, T_n^2]\times S^1} (1 + |\nabla_{g_n}\phi_{\alpha_n}|^2)^{\alpha_n - 2} \langle \nabla_{g_n}\phi_{\alpha_n}, \nabla_{g_n} V_n\rangle^2 dvol_{g_n} \\
&= \mathbb{I} + \mathbb{III}.
\end{aligned}$$

Firstly, it is easy to see that

$$\langle \nabla_{g_n}\phi_{\alpha_n}, \nabla_{g_n} V_n\rangle^2 \le C|\nabla_{g_n}\phi_{\alpha_n}|^4.$$

Since $\Xi = \infty$, then

$$\frac{1}{\sqrt{|\Xi_n|}} = (T_n^2 - T_n^1)o(n),$$

which implies that

$$\frac{\mathbb{III}}{\sqrt{\Xi_n}} \le C\frac{\alpha_n - 1}{\sqrt{|\Xi_n|}} \int_{[T_n^1, T_n^2]\times S^1} (1 + |\nabla_{g_n}\phi_{\alpha_n}|^2)^{\alpha_n} dvol_{g_n} \le C\frac{\alpha_n - 1}{l_n} \to 0,$$

where the last convergence follows from (2.58).

To estimate the first term $\mathbb{I}$, by a direct computation, we obtain

$$\begin{aligned}
\mathbb{I} &= 2\alpha_n \int_{[T_n^1, T_n^2]\times S^1} (1+|\nabla_{g_n}\phi_{\alpha_n}|^2)^{(\alpha_n-1)} \left(\langle \nabla V_n, \nabla V_n\rangle \right.\\
&\quad \left. - R(V_n, \nabla\phi_{\alpha_n}, \nabla\phi_{\alpha_n}, V_n)\right) dtd\theta \\
&= 2\alpha_n \int_{[T_n^1+\lambda(T_n^2-T_n^1), t_n^a]\times S^1} (1+|\nabla_{g_n}\phi_{\alpha_n}|^2)^{(\alpha_n-1)} \\
&\quad \times \left(\langle \nabla_{\frac{\partial\phi_{\alpha_n}}{\partial t}} V_n, \nabla_{\frac{\partial\phi_{\alpha_n}}{\partial t}} V_n\rangle - R(V_n, \frac{\partial\phi_{\alpha_n}}{\partial t}, \frac{\partial\phi_{\alpha_n}}{\partial t}, V_n)\right) dtd\theta \\
&\quad + 2\alpha_n \int_{[T_n^1+\lambda(T_n^2-T_n^1), t_n^a]\times S^1} (1+|\nabla_{g_n}\phi_{\alpha_n}|^2)^{(\alpha_n-1)} \\
&\quad \times \left(\langle \nabla_{\frac{\partial\phi_{\alpha_n}}{\partial \theta}} V_n, \nabla_{\frac{\partial\phi_{\alpha_n}}{\partial \theta}} V_n\rangle - R(V_n, \frac{\partial\phi_{\alpha_n}}{\partial \theta}, \frac{\partial\phi_{\alpha_n}}{\partial \theta}, V_n)\right) dtd\theta \\
&:= \mathbb{I}_1 + \mathbb{I}_2.
\end{aligned}$$

Noting that (2.58) implies

$$\lim_{n\to\infty} (l_n)^{2(1-\alpha_n)} = 1$$

and the fact

$$\|\nabla\phi_{\alpha_n}\|_{C^0([T_n^1,T_n^2]\times S^1)} \le C \sup_{t\in[T_n^1,T_n^2]} E(\phi_{\alpha_n}, [t-1,t+1]\times S^1) \le C\epsilon,$$

we have

$$\lim_{n\to\infty} (1+|\nabla_{g_n}\phi_{\alpha_n}|^2)^{(\alpha_n-1)} = 1.$$

Then by Lemma 2.15 and Lemma 2.16, we get

$$\begin{aligned}
\lim_{n\to\infty} \frac{\mathbb{I}_1}{\sqrt{|\Xi_n|}} &= \lim_{n\to\infty} \frac{2}{\sqrt{|\Xi_n|}} \int_{[T_n^1+\lambda(T_n^2-T_n^1), t_n^a]\times S^1} \\
&\quad \times \left(\langle \nabla_{\frac{\partial\phi_{\alpha_n}}{\partial t}} V_n, \nabla_{\frac{\partial\phi_{\alpha_n}}{\partial t}} V_n\rangle - R(V_n, \frac{\partial\phi_{\alpha_n}}{\partial t}, \frac{\partial\phi_{\alpha_n}}{\partial t}, V_n)\right) dtd\theta \\
&= \lim_{n\to\infty} 2\int_{[0,a]\times S^1} \left(\langle \nabla_{\frac{\partial\phi_{\alpha_n}}{\partial s}} V_n, \nabla_{\frac{\partial\phi_{\alpha_n}}{\partial s}} V_n\rangle \right. \\
&\quad \left. - R(V_n, \frac{\partial\phi_{\alpha_n}}{\partial s}, \frac{\partial\phi_{\alpha_n}}{\partial s}, V_n)\right) \frac{|\frac{ds}{dt}|}{\sqrt{|\Xi_n|}} dsd\theta \\
&= 2\sqrt{\pi}\ I_\gamma(V_0, V_0).
\end{aligned}$$

For the term $\mathbb{I}_2$, we have

$$\begin{aligned}
\frac{\mathbb{I}_2}{\sqrt{|\Xi_n|}} &\le \frac{C}{\sqrt{|\Xi_n|}} \int_0^{2\pi} \int_{T_n^1+\lambda(T_n^2-T_n^1)}^{t_n^a} \\
&\quad \times \left(\langle \nabla_{\frac{\partial\phi_{\alpha_n}}{\partial \theta}} V_n, \nabla_{\frac{\partial\phi_{\alpha_n}}{\partial \theta}} V_n\rangle - R(V_n, \frac{\partial\phi_{\alpha_n}}{\partial \theta}, \frac{\partial\phi_{\alpha_n}}{\partial \theta}, V_n)\right) dtd\theta \\
&\le \frac{C}{\sqrt{|\Xi_n|}} \int_0^{2\pi} \int_{T_n^1+\lambda(T_n^2-T_n^1)}^{t_n^a} |\frac{\partial\phi_{\alpha_n}}{\partial\theta}|^2 dtd\theta.
\end{aligned}$$

For any given $T > 0$, we set

$$m_n = \left[\frac{t_n^a - (T_n^1 + \lambda(T_n^2 - T_n^1))}{T} \right] + 1,$$

which tends to infinity as $n \to \infty$. By (2.61), there holds

$$\sqrt{|\Xi_n|}\, m_n \leq C(a, T).$$

Hence, it follows that

$$\begin{aligned}
\frac{\mathbb{I}_2}{\sqrt{|\Xi_n|}} &\leq \frac{C}{\sqrt{|\Xi_n|}} \int_{\cup_{i=0}^{m_n}[iT,(i+1)T]\times S^1} |\frac{\partial \phi_{\alpha_n}}{\partial \theta}|^2 dt d\theta \\
&\leq \frac{C m_n \sqrt{|\Xi_n|}}{|\Xi_n|} \frac{1}{m_n} \int_{\cup_{i=0}^{m_n}[iT,(i+1)T]\times S^1} |\frac{\partial \phi_{\alpha_n}}{\partial \theta}|^2 dt d\theta \\
&\leq \frac{C(T)}{m_n} \left(\frac{1}{|\Xi_n|} \int_{\cup_{i=0}^{m_n}[iT,(i+1)T]\times S^1} |\frac{\partial \phi_{\alpha_n}}{\partial \theta}|^2 dt d\theta \right).
\end{aligned}$$

In view of Lemma 2.12, we conclude that

$$\lim_{n\to\infty} \frac{\mathbb{I}_2}{\sqrt{|\Xi_n|}} = 0.$$

Combining these together, we get (2.63). Hence, for n large enough, we have the desired inequality

$$\delta^2 E_{\alpha_n}(\phi_{\alpha_n})(V_n, V_n) < 0.$$

Thus, we complete the proof of this lemma. □

Proof of Proposition 2.17. We shall divide the proof into two steps corresponding to the two assumptions in the proposition.

STEP 1. the Ricci curvature of the target manifold (N, g_N) has a positive lower bound.

Since $\mathrm{Ric}_N \geq \lambda_N > 0$, by the well-known Myers's theorem, we know that, if

$$a \geq \frac{\pi}{\sqrt{\lambda_N (n-1)^{-1}}} + 2\epsilon,$$

then there exists a geodesic γ and a tangent vector field $V_0(s)$ on N, which is smooth on γ, and is vanishing on $\gamma|_{[0,\epsilon]}$ and $\gamma|_{[a-\epsilon,a]}$, such that the second variation of length of γ satisfies

$$I_\gamma(V_0, V_0) = \int_0^a (\langle \nabla_{\dot\gamma} V_0, \nabla_{\dot\gamma} V_0 \rangle - R(V_0, \dot\gamma, \dot\gamma, V_0))\, ds < -\delta < 0. \tag{2.64}$$

Theorem 2.2 tells us that the limit of the neck contains a geodesic of infinite length. By (2.60), we can always choose a suitable positive constant $t_n^1 > 0$ such that, when n is large enough, the arc length a of $\phi_n^*(t)$ on $[T_n^1 + \lambda(T_n^2 - T_n^1), T_n^1 + \lambda(T_n^2 - T_n^1) + t_n^1]$ satisfies

$$a > l_N = \frac{\pi}{\sqrt{\lambda_N (n-1)^{-1}}}.$$

Therefore, by Lemma 2.18, there exists a section V_n^1 of $\phi_{\alpha_n}^*(TN)$, which is 0 outside the region $[T_n^1 + \lambda(T_n^2 - T_n^1), T_n^1 + \lambda(T_n^2 - T_n^1) + t_n^1] \times S^1$, satisfying

$$\delta^2 E_{\alpha_n}(\phi_{\alpha_n})(V_n^1, V_n^1) < 0.$$

By the same method, for $t_1 = T_n^1 + \lambda(T_n^2 - T_n^1) + t_n^1$, we can also pick $t_n^2 > 0$ and construct a section V_n^2, which is 0 outside $[t_1, t_1 + t_n^2] \times S^1$, such that

$$\delta^2 E_{\alpha_n}(\phi_{\alpha_n})(V_n^2, V_n^2) < 0.$$

Since the limit of the neck contains a geodesic of infinite length, then, when n is sufficiently large, there exist i_n with $i_n \to \infty$ such that we can construct via the same way as above a series of sections $\{V_n^3, V_n^4, \cdots, V_n^{i_n}\}$ satisfying that for any $1 \le i \le i_n$ there holds

$$\delta^2 E_{\alpha_n}(\phi_{\alpha_n})(V_n^i, V_n^i) < 0.$$

Obviously, V_n^1, V_n^2, $\cdots$, $V_n^{i_n}$ are linearly independent. This means that the Morse index of ϕ_{α_n} tends to infinity as $n \to \infty$.

STEP 2. the target manifold (N, g_N) has a finite fundamental group.

We prove this proposition by contradiction. If it is false, then there exists an uniform integer $\overline{m}$ such that Index $(\phi_{\alpha_n}; E_{\alpha_n}) \le \overline{m} < \infty$. Let $s = s(r)$ be the arc-length parameter as before with $s(T_n^1 + \lambda(T_n^2 - T_n^1)) = 0$, $s(t_n^a) = a$ and $s(s_n^a) = -a$ where $T_n^1 < s_n^a < T_n^1 + \lambda(T_n^2 - T_n^1) < t_n^a < T_n^2$. By Theorem 2.2, $\omega_\alpha(s)$ converges to γ in $C^1([-a, a])$.

Next, we show that the Morse index of γ on any subinterval $[-a, a]$ is uniformly bounded.

In fact, let $V_1(s), \ldots, V_m(s)$ be the m linearly independent tangent vector fields on N, which are smooth on γ, and are vanishing on $\gamma|_{[-a,-a+\epsilon]}$ and $\gamma|_{[a-\epsilon,a]}$ for any big $a > 0$ and small $\epsilon > 0$, such that for any $q \in \{1, \ldots, m\}$, the second variation of the length of γ satisfies

$$I_\gamma(V_q, V_q) = \int_{-b}^{b} (\langle \nabla_{\dot\gamma} V_q, \nabla_{\dot\gamma} V_q \rangle - R(V_q, \dot\gamma, \dot\gamma, V_q)) ds < -\delta < 0. \tag{2.65}$$

Let P be projection from $T\mathbb{R}^K$ to TN. We define

$$V_n^q(s, \theta) = P_{\phi_{\alpha_n}(s,\theta)}(V_q(s)).$$

Similarly to the proof of Lemma 2.18, we have

$$\delta^2 E_{\alpha_n}(\phi_{\alpha_n})(V_n^q, V_n^q) < 0,$$

when $\alpha_n - 1$ is small enough.

Thus,

$$m \le \text{ Index } (\phi_{\alpha_n}; E_{\alpha_n}) \le \overline{m} < \infty,$$

which implies that the Morse index of geodesic γ is uniformly bounded by $\overline{m}$ since a is arbitrary.

Since the target manifold N has a finite fundamental group, by Gromov estimates (see Corollary 3.3.5 in [**20**]), the length of the geodesic is controlled by the bound of the Morse index, i.e.

$$length(\gamma) \le C(Index(\gamma) + 1) \le C(\overline{m} + 1) < \infty,$$

which is a contradiction to $\nu = \infty$. We finished the proof of this proposition. $\square$

At the end of this section, we prove Theorem 2.3.

Proof of Theorem 2.3. It is easy to see that the first conclusion follows from Theorem 2.2 and Proposition 2.17, while the second conclusion follows from Theorem 2.1 and equality

$$\lim_{n\to\infty} 2\kappa_n(\phi_{\alpha_n})|T_n^2 - T_n^1| = 0$$

which is implied from the fact that $\Xi < \infty$. □

CHAPTER 3

Asymptotic analysis of second type neck

In this section, we will study the asymptotic behaviour of the second type necks in the limit process for a sequence of α-harmonic maps as described in Section 1.

Let

$$x_n := (t_n, \theta_n) \in Q_n = [-T_n, T_n] \times S^1$$

be an energy concentration point for α_n-harmonic map sequence, i.e.

$$\limsup_{n\to\infty} E(\phi_{\alpha_n}, D_r(x_n)) \geq \epsilon_0, \ \forall\, r > 0.$$

Here, $D_r(x_n)$ is a small neighborhood of x_n in Q_n which is defined by

$$D_r(x_n) := \{(t,\theta) \in Q_n |\ |t - t_n|^2 + |\theta - \theta_n|^2 \leq r^2\}.$$

Through a translation, we just need to consider that ϕ_{α_n} is a sequence of α_n-harmonic maps defined on

$$D_{r_0}(0) := \{(x^1, x^2) \in \mathbb{R}^2 \mid |x^1|^2 + |x^2|^2 \leq r_0^2\}$$

with Riemannian metric g_n and with uniformly bounded energy E_{α_n} and 0 is the only energy concentration point in $D_{r_0}(0)$, i.e.

$$\sup_n E_{\alpha_n}(\phi_{\alpha_n}, D_{r_0}(0)) \leq \Lambda, \ \ \limsup_{n\to\infty} E(\phi_{\alpha_n}, D_r(0)) \geq \epsilon_0, \ \forall\, r > 0,$$

where

$$g_n(x) = \left(\frac{l_n}{2\pi\cos(\frac{l_n}{2\pi}(t_n + x^1))}\right)^2 ((dx^1)^2 + (dx^2)^2)$$

and $x = (x^1, x^2)$ is the coordinate in $D_{r_0}(0)$.

Since in general multiple bubbles can split off at a blow-up point, similarly to the interior case in [**14**], we shall consider the following more general α_n-energy functionals

$$\widetilde{E}_{\alpha_n,\sigma_n}(\phi_{\alpha_n}) = \int_{D_{r_0}(0)} (\sigma_n + |\nabla_{\widetilde{g}_n}\phi_{\alpha_n}|^2)^{\alpha_n} dvol_{\widetilde{g}_n}, \ \alpha_n > 1,$$

where

(3.1)

$$\widetilde{g}_n = \widetilde{\rho}_n^2\left((dx^1)^2 + (dx^2)^2\right) = \left(\frac{l_n}{2\pi\cos(\frac{l_n}{2\pi}(t_n + \sqrt{\sigma_n}x^1))}\right)^2 ((dx^1)^2 + (dx^2)^2)$$

and $0 < \sigma_n \leq 1$ satisfies

$$0 < \beta_0 \leq \liminf_{n\to\infty} (\sigma_n)^{\alpha_n - 1} \leq 1.$$

Critical points of $\widetilde{E}_{\alpha_n,\sigma_n}$ are also called α_n*-harmonic maps*, which satisfy the following Euler-Lagrange equation

$$\Delta_{\widetilde{g}_n}\phi_{\alpha_n} + (\alpha_n - 1)\frac{\nabla_{\widetilde{g}_n}|\nabla_{\widetilde{g}_n}\phi_{\alpha_n}|^2\nabla_{\widetilde{g}_n}\phi_{\alpha_n}}{\sigma_n + |\nabla_{\widetilde{g}_n}\phi_{\alpha_n}|^2} + A(\phi_{\alpha_n})(\nabla_{\widetilde{g}_n}\phi_{\alpha_n}, \nabla_{\widetilde{g}_n}\phi_{\alpha_n}) = 0. \tag{3.2}$$

It is easy to see that the equation (3.2) has the following two equivalent forms:

$$\Delta\phi_{\alpha_n} + (\alpha_n - 1)\frac{\nabla|\nabla_{\widetilde{g}_n}\phi_{\alpha_n}|^2\nabla\phi_{\alpha_n}}{\sigma_n + |\nabla_{\widetilde{g}_n}\phi_{\alpha_n}|^2} + A(\phi_{\alpha_n})(\nabla\phi_{\alpha_n}, \nabla\phi_{\alpha_n}) = 0, \tag{3.3}$$

$$\begin{aligned} &\mathrm{div}\ \left\{(\sigma_n + |\nabla_{\widetilde{g}_n}\phi_{\alpha_n}|^2)^{\alpha_n-1}\nabla\phi_{\alpha_n}\right\} \\ &+(\sigma_n + |\nabla_{\widetilde{g}_n}\phi_{\alpha_n}|^2)^{\alpha_n-1}A(\phi_{\alpha_n})(\nabla\phi_{\alpha_n}, \nabla\phi_{\alpha_n}) = 0. \end{aligned} \tag{3.4}$$

To present the results in this section, we shall first give a general description of the blow-up procedure and the bubbling phenomena. We refer to [**2**, **14**, **23**] for details on some well known results.

Given a sequence of α_n-harmonic maps $\{\phi_{\alpha_n}\} : (D_{r_0}(0), \widetilde{g}_n) \to N$ with uniformly bounded energy

$$\widetilde{E}_{\alpha_n,\sigma_n}(\phi_{\alpha_n}) \le \Lambda,$$

assuming $\{0\}$ is the only energy concentration point, with the help of small energy regularity Lemma 3.4 and applying similar arguments as in Sacks-Uhlenbeck's scheme, by passing to a subsequence, we know that ϕ_{α_n} converges to some limit harmonic map $\phi : D_{r_0}(0) \to N$ smoothly away from $\{0\}$ as $n \to \infty$. We may also assume there are I bubbles occurring at this point, i.e. there are a sequence of points $\{x^i_{\alpha_n}\}$, $i = 1, \dots, I$, and a sequence of positive numbers $\{\lambda^i_{\alpha_n}\}$ with $x^i_{\alpha_n} \to 0$, $\lambda^i_{\alpha_n} \to 0$ as $n \to \infty$ and one of the following two alternatives holds true:

if $1 \le j_1, j_2 \le I$ and $j_1 \ne j_2$,

(H1) for any fixed $R > 0$, $D_{R\lambda^{j_1}_{\alpha_n}}(x^{j_1}_{\alpha_n}) \cap D_{R\lambda^{j_2}_{\alpha_n}}(x^{j_2}_{\alpha_n}) = \emptyset$, whenever n is sufficiently large.

(H2) $\frac{\lambda^{j_1}_{\alpha_n}}{\lambda^{j_2}_{\alpha_n}} + \frac{\lambda^{j_2}_{\alpha_n}}{\lambda^{j_1}_{\alpha_n}} = \infty$, as $n \to \infty$.

Moreover, the rescaled maps

$$u^i_n(x) := \phi_{\alpha_n}(x^i_{\alpha_n} + \lambda^i_{\alpha_n}x)$$

converge in C^k_{loc} to a nontrivial harmonic map u^i away from finitely many points. Usually, we call the above harmonic map u^i (or the corresponding scaling factors $(x^i_{\alpha_n}, \lambda^i_{\alpha_n})$) a bubble.

To study the asymptotic behavior of the necks, we shall consider the following quantities:

$$\mu_i = \liminf_{n\to\infty}\,(\lambda^i_{\alpha_n})^{2-2\alpha_n}, \quad l_\infty = \liminf_{n\to\infty}\,(\widetilde{l}_n)^{2-2\alpha_n}, \quad \nu_i = \liminf_{n\to\infty}\,(\lambda^i_{\alpha_n})^{-\sqrt{\alpha_n-1}}. \tag{3.5}$$

where

$$\widetilde{l}_n = \frac{l_n}{2\pi\cos(\frac{l_n}{2\pi}t_n)}.$$

It is easy to see that $\nu_i \in [0, \infty]$. Also, we can see that there exists two positive constants l_{max}, $\mu_{max} \ge 1$ such that

$$l_\infty \in [1, l_{max}] \quad \textit{and} \quad \mu_i \in [1, \mu_{max}].$$

The proof is similar to Remark 1.2 in [**14**]. In fact, by (H1) or (H2), without loss of generality we assume $\lambda^1_{\alpha_n}$ is the smallest one, i.e. $\frac{\lambda^1_{\alpha_n}}{\lambda^j_{\alpha_n}} \le C < \infty$ for all $j = 2, \ldots, I$, as $n \to \infty$. We just need to show that

$$\mu_1 = \liminf_{n\to\infty} (\lambda^1_{\alpha_n})^{2-2\alpha_n} \le \mu_{max} \quad and \quad l_\infty \in [1, l_{max}].$$

By classical blow-up theory, we have:

$$u^1_n(x) = \phi_{\alpha_n}(x^1_{\alpha_n} + \lambda^1_{\alpha_n} x) \to u^1 \ in \ C^k_{loc}(\mathbb{R}^2),$$

where $u^1 : \mathbb{R}^2 \to N$ is a nontrivial smooth harmonic map. Therefore, we get

$$\begin{aligned}\Lambda &\ge \lim_{R\to\infty}\lim_{n\to\infty} \int_{D_{\lambda^1_{\alpha_n}R}(x^1_{\alpha_n})} |\nabla_{\widetilde{g}_n}\phi_{\alpha_n}|^{2\alpha_n} dvol_{\widetilde{g}_n} \\ &= \lim_{R\to\infty}\lim_{n\to\infty} (\lambda^1_{\alpha_n}\widetilde{l}_n)^{2-2\alpha_n} \int_{D_R(0)} |\nabla u^1_n|^2 dx \\ &= \mu_1 l_\infty E(u^1).\end{aligned}$$

Since u^1 is a nontrivial harmonic sphere, by energy gap theorem, we have

$$\max\{\mu_1,\ l_\infty\} \le \mu_1 l_\infty \le \frac{\Lambda}{E(u^1)},$$

since $\mu_1,\ l_\infty \ge 1$.

3.1. Some results on the second type neck

In this subsection, we will state some theorems on the second type neck, which will be proved later. We start with the following theorem about a generalized energy identity.

THEOREM 3.1. *Let $(D(0), \widetilde{g}_n) \subset \mathbb{R}^2$ be the unit disk equipped with a metric as in* (3.1)

$$\widetilde{g}_n = \widetilde{\rho}^2_n \left((dx^1)^2 + (dx^2)^2\right).$$

Let $\phi_{\alpha_n} \in C^\infty(D(0), N)$ be a sequence of α_n-harmonic maps satisfying

(a) $\sup_n \widetilde{E}_{\alpha_n,\sigma_n}(\phi_{\alpha_n}) \le \Lambda$, *and* $0 < \beta_0 \le \lim_{n\to\infty}(\sigma_n)^{\alpha_n - 1} \le 1$,

(b) $\phi_{\alpha_n} \to \phi$ *strongly in* $C^\infty_{loc}(D \setminus \{0\}, \mathbb{R}^N)$ *as* $n \to \infty$.

Then there exist a subsequence of ϕ_{α_n} (still denoted by ϕ_{α_n}) and a nonnegative integer I such that, for any $i = 1, \ldots, I$, there exist a point $x^i_{\alpha_n}$, positive numbers $\lambda^i_{\alpha_n}$ and a nonconstant harmonic sphere u^i such that:

(i) $x^i_{\alpha_n} \to 0,\ \lambda^i_{\alpha_n} \to 0,$ *as* $n \searrow \infty$;

(ii) $\lim_{n\to\infty}\left(\frac{\lambda^i_{\alpha_n}}{\lambda^j_{\alpha_n}} + \frac{\lambda^j_{\alpha_n}}{\lambda^i_{\alpha_n}} + \frac{|x^i_{\alpha_n} - x^j_{\alpha_n}|}{\lambda^i_{\alpha_n} + \lambda^j_{\alpha_n}}\right) = \infty$ *for any* $i \ne j$;

(iii) u^i *is the weak limit of* $\phi_{\alpha_n}(x^i_{\alpha_n} + \lambda^i_{\alpha_n} x)$ *in* $W^{1,2}_{loc}(\mathbb{R}^2)$;

(iv) ***Energy identity:*** *we have*

$$\lim_{\delta\to 0}\lim_{n\to\infty} \widetilde{E}_{\alpha_n,\sigma_n}(\phi_{\alpha_n}, D_\delta(0)) = \sum_{i=1}^I l_\infty \mu_i^2 E(w^i), \tag{3.6}$$

where $\mu_i = \lim_{n\to\infty}(\lambda^i_{\alpha_n})^{2-2\alpha_n}$ *and* $l_\infty = \lim_{n\to\infty}(\widetilde{l}_n)^{2-2\alpha_n}$.

The second main theorem in this section is studying the limit of the second type neck, which is a geodesic of the target manifold. Precisely, we will prove

THEOREM 3.2. *Under the assumptions of Theorem 3.1, suppose that there is only one bubble w^1, which is a harmonic sphere, letting $\nu = \lim_{n\to\infty}(\lambda_n)^{-\sqrt{\alpha_n-1}}$, then we have*

(1) *when $\nu = 1$, the set $\phi(D_1(0)) \cup w^1(S^2)$ is a connected set in N;*
(2) *when $\nu \in (1,\infty)$, the set $\phi(D_1(0))$ and $w^1(S^2)$ are connected by a geodesic γ in N with length*

$$L(\gamma) = \sqrt{\frac{E(w^1)}{\pi}}\log\nu;$$

(3) *when $\nu = \infty$, the neck contains at least an infinite length geodesic curve.*

Our third main theorem on the second type neck is to exploit some geometric conditions to ensure that the energy identity holds.

THEOREM 3.3. *Under the assumption of Theorem 3.1, suppose that the sequence $\{\phi_{\alpha_n}\}$ has bounded index and assume that one of the following two assumptions holds:*

(i) *the Ricci curvature of the target manifold (N, g_N) has a positive lower bound, i.e. there exists a positive constant $\lambda_N > 0$ such that $Ric_N \geq \lambda_N > 0$;*
(ii) *the target manifold (N, g_N) has a finite fundamental group.*

Then the limiting necks consist of some finite length geodesics.

Moreover, if

$$l_\infty = \lim_{n\to\infty}(\widetilde{l}_n)^{2-2\alpha_n} = 1,$$

then the energy identity holds.

The rest part of Chapter 3 is organized as follows. In Section 3.2, we will introduce some basic lemmas on the second type neck, such as small energy regularity and Pohozaev type identity. In Section 3.3, we will establish a generalized energy identity, i.e. Theorem 3.1. In Section 3.4, we will study some refined neck analysis on the second type neck and prove Theorem 3.2. The proof of Theorem 3.3 will be given in Section 3.5.

3.2. Some basic lemmas

In this subsection, we will prove two basic lemmas, small energy regularity lemma and Pohozaev type identity.

LEMMA 3.4 (Small energy regularity). *Let $(D, \widetilde{g}_n)$ be an unit disk in $\mathbb{R}^2$ equipped with a metric as in* (3.1)

$$\widetilde{g}_n = \widetilde{\rho}_n^2\left((dx^1)^2 + (dx^2)^2\right).$$

There exist two constants $\epsilon_0' > 0$ and $\alpha_0' > 1$ depending only on N, such that if $\phi_{\alpha_n} : (D_1(0), \widetilde{g}_n) \to N$ is an α_n-harmonic map with $1 < \alpha_n \leq \alpha_0'$ and

$$E(\phi_{\alpha_n}) \leq \epsilon_0',$$

then there holds

$$\|\nabla\phi_{\alpha_n}\|_{C^1(D_{1/2}(0))} \leq C(N)\|\nabla\phi_{\alpha_n}\|_{L^2(D_1(0))}.$$

PROOF. According to (3.3), we know that ϕ_{α_n} satisfies following equation

$$\begin{aligned}\Delta\phi_{\alpha_n} &= -A(\nabla\phi_{\alpha_n},\nabla\phi_{\alpha_n}) - (\alpha_n-1)\frac{\nabla|\nabla_{\widetilde{g}_n}\phi_{\alpha_n}|^2\nabla\phi_{\alpha_n}}{\sigma_n+|\nabla_{\widetilde{g}_n}\phi_{\alpha_n}|^2}\\ &= -A(\nabla\phi_{\alpha_n},\nabla\phi_{\alpha_n}) - (\alpha_n-1)\frac{\nabla(\frac{1}{\widetilde{\rho}_n^2}|\nabla\phi_{\alpha_n}|^2)\nabla\phi_{\alpha_n}}{\sigma_n+\frac{1}{\widetilde{\rho}_n^2}|\nabla\phi_{\alpha_n}|^2}\\ &= -A(\nabla\phi_{\alpha_n},\nabla\phi_{\alpha_n}) - (\alpha_n-1)\frac{\frac{1}{\widetilde{\rho}_n^2}\nabla|\nabla\phi_{\alpha_n}|^2\nabla\phi_{\alpha_n}}{\sigma_n+\frac{1}{\widetilde{\rho}_n^2}|\nabla\phi_{\alpha_n}|^2}\\ &\quad - (\alpha_n-1)\frac{\nabla(\frac{1}{\widetilde{\rho}_n^2})|\nabla\phi_{\alpha_n}|^2\nabla\phi_{\alpha_n}}{\sigma_n+\frac{1}{\widetilde{\rho}_n^2}|\nabla\phi_{\alpha_n}|^2},\end{aligned} \tag{3.7}$$

where

$$\left|(\alpha_n-1)\frac{\frac{1}{\widetilde{\rho}_n^2}\nabla|\nabla\phi_{\alpha_n}|^2\nabla\phi_{\alpha_n}}{\sigma_n+\frac{1}{\widetilde{\rho}_n^2}|\nabla\phi_{\alpha_n}|^2}\right| \leq C(\alpha_n-1)|\nabla^2\phi_{\alpha_n}| \tag{3.8}$$

and

$$\begin{aligned}&\left|(\alpha_n-1)\frac{\nabla(\frac{1}{\widetilde{\rho}_n^2})|\nabla\phi_{\alpha_n}|^2\nabla\phi_{\alpha_n}}{\sigma_n+\frac{1}{\widetilde{\rho}_n^2}|\nabla\phi_{\alpha_n}|^2}\right|\\ &= \left|(\alpha_n-1)\frac{2\widetilde{\rho}_n^{-2}\tan(\frac{l_n}{2\pi}(t_n+\sqrt{\sigma_n}x^1))\frac{l_n}{2\pi}\sqrt{\sigma_n}|\nabla\phi_{\alpha_n}|^2\frac{\partial\phi_{\alpha_n}}{\partial x^1}}{\sigma_n+\frac{1}{\widetilde{\rho}_n^2}|\nabla\phi_{\alpha_n}|^2}\right|\\ &\leq C(\alpha_n-1)|l_n\tan(\frac{l_n}{2\pi}(t_n+\sqrt{\sigma_n}x^1))||\frac{\partial\phi_{\alpha_n}}{\partial x^1}|\\ &\leq C(\alpha_n-1)|\frac{\partial\phi_{\alpha_n}}{\partial x^1}|\end{aligned} \tag{3.9}$$

where we used the fact that

$$\frac{d\widetilde{\rho}_n^{-2}}{dx^1} = -2\widetilde{\rho}_n^{-2}\tan(\frac{l_n}{2\pi}(t_n+\sqrt{\sigma_n}x^1))\frac{l_n}{2\pi}\sqrt{\sigma_n}.$$

The rest of the proof is similar to the argument in Lemma 2.5, or one can also see [**23**]. □

Next, we derive a Pohozaev type identity by using a method of integrating by parts.

LEMMA 3.5. *Let $(D,\widetilde{g}_n)$ be an unit disk in $\mathbb{R}^2$ equipped with a metric as in* (3.1)

$$\widetilde{g}_n = \widetilde{\rho}_n^2\left((dx^1)^2+(dx^2)^2\right).$$

If ϕ_{α_n} is a critical point of $E_{\alpha_n,\sigma_n}(\phi_{\alpha_n}, D)$, then for any $0 < t < 1$, there holds

$$\begin{aligned}
&(1-\frac{1}{2\alpha_n})\int_{\partial D_t}(\sigma_n+|\nabla_{\widetilde{g}_n}\phi_{\alpha_n}|^2)^{\alpha_n-1}|\frac{\partial\phi_{\alpha_n}}{\partial r}|^2\\
&\quad-\frac{1}{2\alpha_n}\int_{\partial D_t}(\sigma_n+|\nabla_{\widetilde{g}_n}\phi_{\alpha_n}|^2)^{\alpha_n-1}|x|^{-2}|\frac{\partial\phi_{\alpha_n}}{\partial\theta}|^2\\
&\quad=(1-\frac{1}{\alpha_n})\frac{1}{t}\int_{D_t}(\sigma_n+|\nabla_{\widetilde{g}_n}\phi_{\alpha_n}|^2)^{\alpha_n-1}|\nabla\phi_{\alpha_n}|^2dx\\
&\quad+\frac{\sigma_n}{2\alpha_n}\int_{\partial D_t}(\sigma_n+|\nabla_{\widetilde{g}_n}\phi_{\alpha_n}|^2)^{\alpha_n-1}\widetilde{\rho}_n^2\\
&\quad+(1-\frac{1}{\alpha_n})\frac{1}{t}\int_{D_t}(\sigma_n+|\nabla_{\widetilde{g}_n}\phi_{\alpha_n}|^2)^{\alpha_n-1}|\nabla\phi_{\alpha_n}|^2\\
&\quad\times\tan(\frac{l_n}{2\pi}(t_n+\sqrt{\sigma_n}x^1))\frac{l_nx^1}{2\pi}\sqrt{\sigma_n}dx\\
&\quad-\frac{\sigma_n}{\alpha_nt}\int_{D_t}(\sigma_n+|\nabla_{\widetilde{g}_n}\phi_{\alpha_n}|^2)^{\alpha_n-1}\widetilde{\rho}_n^2\left(1+\tan(\frac{l_n}{2\pi}(t_n+\sqrt{\sigma_n}x^1))\frac{l_nx^1}{2\pi}\sqrt{\sigma_n}\right)dx,
\end{aligned}\tag{3.10}$$

where (r, θ) is the polar coordinate system at point 0 and $dx = dx^1dx^2$.

PROOF. Multiplying (3.4) by $r\frac{\partial\phi_{\alpha_n}}{\partial r}$ and integrating by parts, we have

$$\begin{aligned}
0&=\int_{D_t}div\{(\sigma_n+|\nabla_{\widetilde{g}_n}\phi_{\alpha_n}|^2)^{\alpha_n-1}\nabla\phi_{\alpha_n}\}r\frac{\partial\phi_{\alpha_n}}{\partial r}dx\\
&=\int_{\partial D_t}(\sigma_\alpha+|\nabla_{\widetilde{g}_n}\phi_{\alpha_n}|^2)^{\alpha_n-1}r|\frac{\partial\phi_{\alpha_n}}{\partial r}|^2\\
&\quad-\int_{D_t}(\sigma_n+|\nabla_{\widetilde{g}_n}\phi_{\alpha_n}|^2)^{\alpha_n-1}\nabla\phi_{\alpha_n}\nabla(x^\beta\frac{\partial\phi_{\alpha_n}}{\partial x^\beta})dx\\
&=\int_{\partial D_t}(\sigma_n+|\nabla_{\widetilde{g}_n}\phi_{\alpha_n}|^2)^{\alpha_n-1}r|\frac{\partial\phi_{\alpha_n}}{\partial r}|^2\\
&\quad-\int_{D_t}(\sigma_n+|\nabla_{\widetilde{g}_n}\phi_{\alpha_n}|^2)^{\alpha_n-1}|\nabla\phi_{\alpha_n}|^2dx\\
&\quad-\int_{D_t}(\sigma_n+|\nabla_{\widetilde{g}_n}\phi_{\alpha_n}|^2)^{\alpha_n-1}\frac{1}{2}x^\beta\frac{\partial}{\partial x^\beta}|\nabla\phi_{\alpha_n}|^2dx.
\end{aligned}\tag{3.11}$$

Noting that

$$|\nabla\phi_{\alpha_n}|^2=\widetilde{\rho}_n^2|\nabla_{\widetilde{g}_n}\phi_{\alpha_n}|^2$$

and

$$\frac{\partial\widetilde{\rho}_n^2}{\partial x^1}=2\widetilde{\rho}_n^2\tan(\frac{l_n}{2\pi}(t_n+\sqrt{\sigma_n}x^1))\frac{l_n}{2\pi}\sqrt{\sigma_n},$$

we get

$$
\begin{aligned}
&-\int_{D_t}(\sigma_n+|\nabla_{\widetilde{g}_n}\phi_{\alpha_n}|^2)^{\alpha_n-1}\frac{1}{2}x^\beta\frac{\partial}{\partial x^\beta}|\nabla\phi_{\alpha_n}|^2dx\\
&=-\int_{D_t}(\sigma_n+|\nabla_{\widetilde{g}_n}\phi_{\alpha_n}|^2)^{\alpha_n-1}\frac{1}{2}x^\beta\frac{\partial}{\partial x^\beta}|\nabla_{\widetilde{g}_n}\phi_{\alpha_n}|^2\widetilde{\rho}_n^2dx\\
&\quad-\int_{D_t}(\sigma_n+|\nabla_{\widetilde{g}_n}\phi_{\alpha_n}|^2)^{\alpha_n-1}|\nabla\phi_{\alpha_n}|^2\tan(\frac{l_n}{2\pi}(t_n+\sqrt{\sigma_n}x^1))\frac{l_nx^1}{2\pi}\sqrt{\sigma_n}dx\\
&=-\frac{1}{2\alpha_n}\int_{D_t}x^\beta\frac{\partial}{\partial x^\beta}(\sigma_n+|\nabla_{\widetilde{g}_n}\phi_{\alpha_n}|^2)^{\alpha_n}\widetilde{\rho}_n^2dx\\
&\quad-\int_{D_t}(\sigma_n+|\nabla_{\widetilde{g}_n}\phi_{\alpha_n}|^2)^{\alpha_n-1}|\nabla\phi_{\alpha_n}|^2\tan(\frac{l_n}{2\pi}(t_n+\sqrt{\sigma_n}x^1))\frac{l_nx^1}{2\pi}\sqrt{\sigma_n}dx.
\end{aligned}\tag{3.12}
$$

Integrating by parts yields that

$$
\begin{aligned}
&-\frac{1}{2\alpha_n}\int_{D_t}x^\beta\frac{\partial}{\partial x^\beta}(\sigma_n+|\nabla_{\widetilde{g}_n}\phi_{\alpha_n}|^2)^{\alpha_n}\widetilde{\rho}_n^2dx\\
&=-\frac{1}{2\alpha_n}\int_{\partial D_t}r(\sigma_n+|\nabla_{\widetilde{g}_n}\phi_{\alpha_n}|^2)^{\alpha_n}\widetilde{\rho}_n^2\\
&\quad+\frac{1}{2\alpha_n}\int_{D_t}(\sigma_n+|\nabla_{\widetilde{g}_n}\phi_{\alpha_n}|^2)^{\alpha_n}div\{x\widetilde{\rho}_n^2\}dx\\
&=-\frac{1}{2\alpha_n}\int_{\partial D_t}r(\sigma_n+|\nabla_{\widetilde{g}_n}\phi_{\alpha_n}|^2)^{\alpha_n-1}|\nabla\phi_{\alpha_n}|^2\\
&\quad-\frac{\sigma_n}{2\alpha_n}\int_{\partial D_t}r(\sigma_n+|\nabla_{\widetilde{g}_n}\phi_{\alpha_n}|^2)^{\alpha_n-1}\widetilde{\rho}_n^2\\
&\quad+\frac{1}{\alpha_n}\int_{D_t}(\sigma_n+|\nabla_{\widetilde{g}_n}\phi_{\alpha_n}|^2)^{\alpha_n}\widetilde{\rho}_n^2dx\\
&\quad+\frac{1}{\alpha_n}\int_{D_t}(\sigma_n+|\nabla_{\widetilde{g}_n}\phi_{\alpha_n}|^2)^{\alpha_n}\widetilde{\rho}_n^2\tan(\frac{l_n}{2\pi}(t_n+\sqrt{\sigma_n}x^1))\frac{l_nx^1}{2\pi}\sqrt{\sigma_n}dx.
\end{aligned}\tag{3.13}
$$

Combining these together, we obtain

$$\begin{aligned}
&\int_{\partial D_t}(\sigma_n+|\nabla_{\tilde{g}_n}\phi_{\alpha_n}|^2)^{\alpha_n-1}r|\frac{\partial\phi_{\alpha_n}}{\partial r}|^2-\frac{1}{2\alpha_n}\int_{\partial D_t}r(\sigma_n+|\nabla_{\tilde{g}_n}\phi_{\alpha_n}|^2)^{\alpha_n-1}|\nabla\phi_{\alpha_n}|^2\\
&=\frac{\sigma_n}{2\alpha_n}\int_{\partial D_t}r(\sigma_n+|\nabla_{\tilde{g}_n}\phi_{\alpha_n}|^2)^{\alpha_n-1}\tilde{\rho}_n^2+\int_{D_t}(\sigma_n+|\nabla_{\tilde{g}_n}\phi_{\alpha_n}|^2)^{\alpha_n-1}|\nabla\phi_{\alpha_n}|^2dx\\
&\quad-\frac{1}{\alpha_n}\int_{D_t}(\sigma_n+|\nabla_{\tilde{g}_n}\phi_{\alpha_n}|^2)^{\alpha_n}\tilde{\rho}_n^2dx\\
&\quad+\int_{D_t}(\sigma_n+|\nabla_{\tilde{g}_n}\phi_{\alpha_n}|^2)^{\alpha_n-1}|\nabla\phi_{\alpha_n}|^2\tan(\frac{l_n}{2\pi}(t_n+\sqrt{\sigma_n}x^1))\frac{l_nx^1}{2\pi}\sqrt{\sigma_n}dx\\
&\quad-\frac{1}{\alpha_n}\int_{D_t}(\sigma_n+|\nabla_{\tilde{g}_n}\phi_{\alpha_n}|^2)^{\alpha_n}\tilde{\rho}_n^2\tan(\frac{l_n}{2\pi}(t_n+\sqrt{\sigma_n}x^1))\frac{l_nx^1}{2\pi}\sqrt{\sigma_n}dx\\
&=\frac{\sigma_n}{2\alpha_n}\int_{\partial D_t}r(\sigma_n+|\nabla_{\tilde{g}_n}\phi_{\alpha_n}|^2)^{\alpha_n-1}\tilde{\rho}_n^2\\
&\quad+(1-\frac{1}{\alpha_n})\int_{D_t}(\sigma_n+|\nabla_{\tilde{g}_n}\phi_{\alpha_n}|^2)^{\alpha_n-1}|\nabla\phi_{\alpha_n}|^2dx\\
&\quad-\frac{\sigma_n}{\alpha_n}\int_{D_t}(\sigma_n+|\nabla_{\tilde{g}_n}\phi_{\alpha_n}|^2)^{\alpha_n-1}\tilde{\rho}_n^2dx\\
&\quad+(1-\frac{1}{\alpha_n})\int_{D_t}(\sigma_n+|\nabla_{\tilde{g}_n}\phi_{\alpha_n}|^2)^{\alpha_n-1}|\nabla\phi_{\alpha_n}|^2\\
&\quad\times\tan(\frac{l_n}{2\pi}(t_n+\sqrt{\sigma_n}x^1))\frac{l_nx^1}{2\pi}\sqrt{\sigma_n}dx\\
&\quad-\frac{\sigma_n}{\alpha_n}\int_{D_t}(\sigma_n+|\nabla_{\tilde{g}_n}\phi_{\alpha_n}|^2)^{\alpha_n-1}\tilde{\rho}_n^2\tan(\frac{l_n}{2\pi}(t_n+\sqrt{\sigma_n}x^1))\frac{l_nx^1}{2\pi}\sqrt{\sigma_n}dx
\end{aligned}$$

which implies the conclusion of the lemma. □

As a direct corollary of above lemma, we have

COROLLARY 3.6. *Let $(D,\tilde{g}_n)$ be a unit disk in $\mathbb{R}^2$ equipped with a metric as in* (3.1)

$$\tilde{g}_n=\tilde{\rho}_n^2\left((dx^1)^2+(dx^2)^2\right).$$

If ϕ_{α_n} is a critical point of $E_{\alpha_n,\sigma_n}(\phi_{\alpha_n},D)$ where $0<\beta_0<\lim_{\alpha\searrow 1}(\sigma_n)^{\alpha_n-1}\leq 1$ and $E_{\alpha_n,\sigma_n}(\phi_{\alpha_n})\leq\Lambda$, then for any $0<t<\frac{1}{2}$, we have

$$\begin{aligned}
&(1-\frac{1}{2\alpha_n})\int_{\partial D_t}(\sigma_n+|\nabla_{\tilde{g}_n}\phi_{\alpha_n}|^2)^{\alpha_n-1}|\frac{\partial\phi_{\alpha_n}}{\partial r}|^2\\
&\quad-\frac{1}{2\alpha_n}\int_{\partial D_t}(\sigma_n+|\nabla_{\tilde{g}_n}\phi_{\alpha_n}|^2)^{\alpha_n-1}|x|^{-2}|\frac{\partial\phi_{\alpha_n}}{\partial\theta}|^2\\
&=(1-\frac{1}{\alpha_n})\frac{1}{t}\int_{D_t}(\sigma_n+|\nabla_{\tilde{g}_n}\phi_{\alpha_n}|^2)^{\alpha_n-1}|\nabla\phi_{\alpha_n}|^2dx+O(t)+O(\alpha_n-1).
\end{aligned}\tag{3.14}$$

PROOF. Without loss of generality, we assume 0 is the only energy concentration point for the sequence $\{\phi_{\alpha_n}\}$. Then by the blow-up process described at the beginning of this section, we can get finitely many bubbles at this point, i.e. there exist a positive sequence $\lambda_{\alpha_n}^i\to 0$ and a sequence of points $x_{\alpha_n}^i\to 0$, $i=1,\dots,I$, as $n\to\infty$, which satisfy (H1) or (H2). Also, without loss of generality, we assume

$\lambda^1_{\alpha_n}$ is the smallest one, i.e. $\limsup_{n\to\infty}\frac{\lambda^1_{\alpha_n}}{\lambda^j_{\alpha_n}} \le C$ for $j = 2, \dots, I$. By a standard blow-up argument, we know $u_n(x) = \phi_{\alpha_n}(x^1_{\alpha_n} + \lambda^1_{\alpha_n} x)$ has no energy concentration points. By Lemma 3.4, we have

$$\|(\sigma_n + |\nabla_{\tilde{g}_n}\phi_{\alpha_n}|^2)^{\alpha_n-1}\|_{C^0(D_{\frac{1}{2}})} \le C(1+(\lambda^1_{\alpha_n}\tilde{l}_n)^{2-2\alpha_n}) \le C(1+\mu_{max}l_{max}). \tag{3.15}$$

Then the conclusion of the lemma follows immediately since

$$|\tilde{\rho}_n| \le C, \quad l_n \tan(\frac{l_n}{2\pi}(t_n + \sqrt{\sigma_0}x^1)) \le C.$$

□

For the sake of convenience, we recall a decay estimate in [**11**] at the end of this subsection, which will be used in next section.

LEMMA 3.7 (Lemma 3.4 and Corollary 3.5 in [**11**]). *Let L be the positive constant in Proposition 2.6. Let $u \in W^{2,2}(D_{e^{(l+1)L}r_2} \setminus D_{r_2})$ for some integer $l > 1$, $f_1 \in L^2(D_{e^{(l+1)L}r_2} \setminus D_{r_2})$, satisfying*

$$\Delta u = A^1 u + A^2 \nabla u + \frac{1}{2\pi}\int_0^{2\pi} A^3 u + A^4 \nabla u d\theta + f_1.$$

Then there exists a positive constant $\delta_0 > 0$ such that if

$$|x|^2(|A^1| + |A^3|) + |x|(|A^2| + |A^4|) \le \delta_0,$$

and

$$\int_{\partial D_r} u = 0, \quad for\ any\ r_2 \le r \le e^{(l+1)L}r_2,$$

then for any $1 \le i \le l$, there holds

$$\begin{aligned}\|\nabla u\|_{L^2(P'_i)} &\le C \max_{j=1,\dots,l} \||x|f_1\|_{L^2(P'_j)} \\ &\quad + C\left(\|\frac{u}{|x|}\|_{L^2(P'_0)} + \|\frac{u}{|x|}\|_{L^2(P'_l)}\right)\left(e^{-\frac{1}{2}(l-i)L} + e^{-\frac{1}{2}iL}\right),\end{aligned}$$

where

$$P'_i := D_{e^{(i+1)L}r_2} \setminus D_{e^{iL}r_2}.$$

3.3. Generalized energy identity

In this subsection, we prove Theorem 3.1. We will use some ideas in [**14**] where they proved a generalized energy identity for a blow-up sequence of α-harmonic maps with fixed domain metric. In fact, the arguments in [**14**] can also be applied to the case of varying domain metrics that are non-degenerating (i.e. the situation involving the third type neck). However, since there are some new quantities appearing and we will use the three circle type method instead of the method of integrating by parts in [**14**], we sketch all the details here.

Proof of Theorem 3.1. By our assumptions, we know that 0 is the only blow-up point of the sequence $\{(\phi_{\alpha_n}\}$ in D, *i.e.*

$$\liminf_{\alpha\searrow 1} E(\phi_{\alpha_n}, D_r) \ge \frac{(\epsilon'_0)^2}{8} \text{ for all } r > 0 \tag{3.16}$$

where $\epsilon_0' > 0$ is the constant in Lemma 3.4. By the standard blow-up argument, we can assume that there exist sequences $x_n \to 0$ and $\lambda_n \to 0$ such that

$$E(\phi_{\alpha_n}, D_{\lambda_n}(x_n)) = \sup_{\substack{x\in D, r\le\lambda_\alpha \\ D_r(x)\subset D}} E(\phi_{\alpha_n}, D_r(x)) = \frac{(\epsilon_0')^2}{32}. \tag{3.17}$$

Set

$$w_n(x) := \phi_{\alpha_n}(x_n + \lambda_n x).$$

It is easy to see that $w_n(x)$ is an α_n-harmonic map living in a region which tends to $\mathbb{R}^2$ as $n \to \infty$. Moreover, for any $x \in \mathbb{R}^2$, when n is sufficiently large, by (3.17), we have

$$E(w_n, D_1(x)) \le \frac{(\epsilon_0')^2}{32}. \tag{3.18}$$

According to Lemma 3.4, there exist a subsequence of w_n (we still denote it by w_n) and a harmonic map $w^1 \in W^{1,2}(\mathbb{R}^2, N)$ such that

$$\lim_{n\to\infty} w_n(x) = w^1(x) \quad \text{in } W^{1,2}_{loc}(\mathbb{R}^2). \tag{3.19}$$

Besides, we know $E(w^1, D_1(0)) = \frac{(\epsilon_0')^2}{32}$. By classical blow-up theory of harmonic maps, w^1 can be extended to a nontrivial harmonic sphere. We call the above nontrivial harmonic sphere w^1 the first bubble.

By a standard induction argument as in the case of approximate harmonic maps with L^2-uniformly bounded tension fields [**5**] or as in the case of α-harmonic maps [**14**], we only need to prove the theorem in the case where there is only one bubble. Under the "one bubble" assumption, we first make the following:

CLAIM. For any $\epsilon > 0$, there exist $\delta > 0$ and $R > 0$ such that

$$\int_{D_{8t}(x_n)\setminus D_t(x_n)} |\nabla\phi_{\alpha_n}|^2 dx \le \epsilon^2 \text{ for any } t \in (\frac{1}{4}\lambda_n R, 2\delta) \tag{3.20}$$

when n is big enough.

The proof of this claim is now standard and it follows from a contradiction argument. In fact, if the claim is not true, one can then get a second bubble. For more details on such a contradiction argument, one can refer to [**2, 5**] for harmonic maps and [**14**] for α-harmonic maps.

To proceed, we need to establish a series of auxiliary lemmas and we leave the rest of the proof of Theorem 3.1 to the end of this section. □

LEMMA 3.8. *Under the assumption of Theorem 3.1, if there is no energy concentration for the sequence ϕ_{α_n} in the region $D_\delta(x_n)\setminus D_{\lambda_n R}(x_n)$, i.e. (3.20) holds, we have for any $\lambda_n R \le t_1 \le t_2 \le \delta$, there holds*

$$\int_{D_{t_2}(x_n)\setminus D_{t_1}(x_n)} |\nabla^2\phi_{\alpha_n}||\phi_{\alpha_n} - \phi^*_{\alpha_n}| dx \le C \int_{D_{2t_2}(x_n)\setminus D_{\frac{1}{2}t_1}(x_n)} |\nabla\phi_{\alpha_n}|^2 dx,$$

where $\phi^*_{\alpha_n}(r) = \frac{1}{2\pi}\int_0^{2\pi} \phi_{\alpha_n}(r,\theta)d\theta$ and $C > 0$ is independent of n.

PROOF. Without loss of generality, we assume $t_2 = 2^I t_1$ for some positive integer I. Under the assumption (3.20) holds which implies

$$|x-x_n||\nabla\phi_{\alpha_n}(x)|\leq C\int_{D_{2^{i+1}t_1}(x_n)\setminus D_{2^{i-2}t_1}(x_n)}|\nabla\phi_{\alpha_n}|^2dx,\ \ 2^{i-1}t_1\leq|x-x_n|\leq 2^i t_1,\quad 1\leq i\leq I,$$

and

$$\int_{D_{2^i t_1}(x_n)\setminus D_{2^{i-1}t_1}(x_n)}|\nabla^2\phi_{\alpha_n}|^2dx\leq C\int_{D_{2^{i+1}t_1}(x_n)\setminus D_{2^{i-2}t_1}(x_n)}|\nabla\phi_{\alpha_n}|^2dx,\quad 1\leq i\leq I$$

by Lemma 3.4.

Then we get

$$\begin{aligned}
&\int_{D_{t_2}(x_n)\setminus D_{t_1}(x_n)}|\nabla^2\phi_{\alpha_n}||x-x_n||\nabla\phi_{\alpha_n}|dx\\
&=\sum_{i=1}^{I}\int_{D_{2^i t_1}(x_n)\setminus D_{2^{i-1}t_1}(x_n)}|\nabla^2\phi_{\alpha_n}||x-x_n||\nabla\phi_{\alpha_n}|dx\\
&\leq\sum_{i=1}^{I}\sup_{2^{i-1}t_1\leq|x-x_n|\leq 2^i t_1}|x-x_n||\nabla\phi_{\alpha_n}|\int_{D_{2^i t_1}(x_n)\setminus D_{2^{i-1}t_1}(x_n)}|\nabla^2\phi_{\alpha_n}|dx\\
&\leq C\sum_{i=1}^{I}\int_{D_{2^{i+1}t_1}(x_n)\setminus D_{2^{i-2}t_1}(x_n)}|\nabla\phi_{\alpha_n}|^2dx\\
&= C\int_{D_{2t_2}(x_n)\setminus D_{t_1/2}(x_n)}|\nabla\phi_{\alpha_n}|^2dx.
\end{aligned}$$

□

To estimate the whole energy on the neck domain, we first prove that the tangential energy is converging to zero.

LEMMA 3.9. *Under the assumptions of Lemma 3.8, there holds:*

$$\lim_{\delta\to 0}\lim_{R\to\infty}\lim_{n\to\infty}\int_{D_\delta(x_n)\setminus D_{\lambda_n R}(x_n)}|x-x_n|^{-2}|\frac{\partial\phi_{\alpha_n}}{\partial\theta}|^2dx=0;$$

PROOF. Denoting

$$\phi^*_{\alpha_n}(r)=\frac{1}{2\pi}\int_0^{2\pi}\phi_{\alpha_n}(r,\theta)d\theta, \tag{3.21}$$

where (r,θ) is the polar coordinate system at point x_n, then by (3.20) and Lemma 3.4, we know that for any $\epsilon>0$, there holds

$$\begin{aligned}
|\phi_{\alpha_n}(r,\theta)-\phi^*_{\alpha_n}(r)|&\leq C\|\nabla\phi_{\alpha_n}(r,\theta)\|_{L^\infty(\{r\}\times S^1)}\\
&\leq CE(\phi_{\alpha_n},D_{2r}(x_n)\setminus D_{\frac{1}{2}r}(x_n))\\
&\leq C\epsilon,\ \forall\ \lambda_n R\leq r\leq\delta,
\end{aligned}$$

when n is big enough.

Integrating by parts and using equation (3.7)-(3.9), we get

$$\begin{aligned}
&\int_{D_\delta(x_n)\setminus D_{\lambda_n R}(x_n)} \nabla\phi_{\alpha_n}\nabla(\phi_{\alpha_n}-\phi^*_{\alpha_n})dx - \int_{\partial D_\delta(x_n)} \frac{\partial\phi_{\alpha_n}}{\partial r}(\phi_{\alpha_n}-\phi^*_{\alpha_n})d\theta \\
&\quad + \int_{\partial D_{\lambda_n R}(x_n)} \frac{\partial\phi_{\alpha_n}}{\partial r}(\phi_{\alpha_n}-\phi^*_{\alpha_n})d\theta \\
&= \int_{D_\delta(x_n)\setminus D_{\lambda_n R}(x_n)} -\Delta\phi_{\alpha_n}(\phi_{\alpha_n}-\phi^*_{\alpha_n})dx \\
&= \int_{[D_\delta(x_n)\setminus D_{\lambda_n R}(x_n)} A(\nabla\phi_{\alpha_n},\nabla\phi_{\alpha_n})(\phi_{\alpha_n}-\phi^*_{\alpha_n})dx \\
&\quad + \int_{D_\delta(x_n)\setminus D_{\lambda_n R}(x_n)} \left((\alpha_n-1)\frac{\frac{1}{\widetilde{\rho}_n^2}\nabla|\nabla\phi_{\alpha_n}|^2\nabla\phi_{\alpha_n}}{1+\frac{1}{\widetilde{\rho}_n^2}|\nabla\phi_{\alpha_n}|^2}\right. \\
&\qquad\qquad \left.+(\alpha_n-1)\frac{\nabla(\frac{1}{\widetilde{\rho}_n^2})|\nabla\phi_{\alpha_n}|^2\nabla\phi_{\alpha_n}}{1+\frac{1}{\widetilde{\rho}_n^2}|\nabla\phi_{\alpha_n}|^2}\right)(\phi_{\alpha_n}-\phi^*_{\alpha_n})dx \\
&\leq C(N)\epsilon\int_{D_\delta(x_n)\setminus D_{\lambda_n R}(x_n)} |\nabla\phi_{\alpha_n}|^2dx \\
&\quad + C(N)(\alpha_n-1)\int_{D_\delta(x_n)\setminus D_{\lambda_n R}(x_n)} (|\nabla^2\phi_{\alpha_n}|+|\nabla\phi_{\alpha_n}|)|\phi_{\alpha_n}-\phi^*_{\alpha_n}|dx.
\end{aligned}$$

On one hand, noting that

$$\begin{aligned}
&\int_{D_\delta(x_n)\setminus D_{\lambda_n R}(x_n)} \nabla\phi_{\alpha_n}\nabla\phi^*_{\alpha_n}dx \\
&\quad = \int_{D_\delta(x_n)\setminus D_{\lambda_n R}(x_n)} \frac{\partial\phi_{\alpha_n}}{\partial r}\frac{\partial\phi^*_{\alpha_n}}{\partial r}dx \\
&\quad \leq \left(\int_{D_\delta(x_n)\setminus D_{\lambda_n R}(x_n)} |\frac{\partial\phi_{\alpha_n}}{\partial r}|^2dx\right)^{\frac{1}{2}}\left(\int_{D_\delta(x_n)\setminus D_{\lambda_n R}(x_n)} |\frac{\partial\phi^*_{\alpha_n}}{\partial r}|^2dx\right)^{\frac{1}{2}} \\
&\quad \leq \int_{D_\delta(x_n)\setminus D_{\lambda_n R}(x_n)} |\frac{\partial\phi_{\alpha_n}}{\partial r}|^2dx,
\end{aligned}$$

we obtain

$$\begin{aligned}
&\int_{D_\delta(x_n)\setminus D_{\lambda_n R}(x_n)} \nabla\phi_{\alpha_n}\nabla(\phi_{\alpha_n}-\phi^*_{\alpha_n})dx \\
&\quad \geq \int_{D_\delta(x_n)\setminus D_{\lambda_n R}(x_n)} |\nabla\phi_{\alpha_n}|^2dx - \int_{D_\delta(x_n)\setminus D_{\lambda_n R}(x_n)} |\frac{\partial\phi_{\alpha_n}}{\partial r}|^2dx \\
&\quad = \int_{D_\delta(x_n)\setminus D_{\lambda_n R}(x_n)} |\frac{\partial\phi_{\alpha_n}}{\partial\theta}|^2dx.
\end{aligned}$$

On the other hand, by Lemma 3.8, we have

$$\int_{D_\delta(x_n)\setminus D_{\lambda_n R}(x_n)} |\nabla^2\phi_{\alpha_n}||\phi_{\alpha_n}-\phi^*_{\alpha_n}|dx \leq CE(\phi_{\alpha_n}, D_{2\delta}(x_n)\setminus D_{\frac{1}{2}\lambda_n R}(x_n)) \leq C\Lambda,$$

then it yields

$$\begin{aligned}
&\int_{D_\delta(x_n)\setminus D_{\lambda_n R}(x_n)} (|\nabla^2\phi_{\alpha_n}| + |\nabla\phi_{\alpha_n}|)|\phi_{\alpha_n} - \phi^*_{\alpha_n}|dx \\
&\leq C\Lambda + \left(\int_{[T_n^1,T_n^2]\times S^1} |\nabla\phi_{\alpha_n}|^2 dx\right)^{\frac{1}{2}} \left(\int_{[T_n^1,T_n^2]\times S^1} |\phi_{\alpha_n} - \phi^*_{\alpha_n}|^2 dx\right)^{\frac{1}{2}} \\
&\leq C\Lambda + C\sqrt{\Lambda}\left(\int_{[T_n^1,T_n^2]\times S^1} |\nabla\phi_{\alpha_n}|^2 dx\right)^{\frac{1}{2}} \leq C\Lambda,
\end{aligned}$$

where we used Poincaré inequality.

Combining these together, we arrived at

$$\begin{aligned}
\int_{D_\delta(x_n)\setminus D_{\lambda_n R}(x_n)} |\frac{\partial\phi_{\alpha_n}}{\partial\theta}|^2 dx \leq C\left(\epsilon + (\alpha_n - 1)\right) + \left|\int_{\partial D_\delta(x_n)} \frac{\partial\phi_{\alpha_n}}{\partial r}(\phi_{\alpha_n} - \phi^*_{\alpha_n})d\theta\right| \\
+ \left|\int_{\partial D_{\lambda_n R}(x_n)} \frac{\partial\phi_{\alpha_n}}{\partial r}(\phi_{\alpha_n} - \phi^*_{\alpha_n})d\theta\right|.
\end{aligned}$$

For the boundary term, by trace theorem and Lemma 3.4, we have

$$\begin{aligned}
\left|\int_{\partial D_\delta(x_n)} \frac{\partial\phi_{\alpha_n}}{\partial r}(\phi_{\alpha_n} - \phi^*_{\alpha_n})d\theta\right| &\leq C\epsilon\int_{\partial D_\delta(x_n)} |\frac{\partial\phi_{\alpha_n}}{\partial r}|d\theta \\
&\leq C\epsilon\|\nabla\phi_{\alpha_n}\|_{W^{1,2}(D_\delta(x_n)\setminus D_{\frac{3}{4}\delta}(x_n))} \\
&\leq C\epsilon\|\nabla\phi_{\alpha_n}\|_{L^2(D_{2\delta}(x_n)\setminus D_{\frac{1}{2}\delta}(x_n))} \leq C\epsilon.
\end{aligned}$$

Similarly, there holds

$$\left|\int_{\partial D_{\lambda_n R}(x_n)} \frac{\partial\phi_{\alpha_n}}{\partial r}(\phi_{\alpha_n} - \phi^*_{\alpha_n})d\theta\right| \leq C\epsilon.$$

Thus, we conclude that

$$\int_{D_\delta(x_n)\setminus D_{\lambda_n R}(x_n)} |\frac{\partial\phi_{\alpha_n}}{\partial\theta}|^2 dtd\theta \leq C\left(\epsilon + (\alpha_n - 1)\right),$$

which implies the conclusion of this lemma. □

By (3.15), it is easy to see that the following corollary holds.

COROLLARY 3.10. *Under the assumptions of Lemma 3.8, there holds:*

$$\lim_{\delta\to 0}\lim_{R\to\infty}\lim_{n\to\infty}\int_{D_\delta(x_n)\setminus D_{\lambda_n R}(x_n)} (\sigma_n + |\nabla_{\widetilde{g}_n}\phi_{\alpha_n}|^2)^{\alpha_n - 1}|x - x_n|^{-2}|\frac{\partial\phi_{\alpha_n}}{\partial\theta}|^2 dx = 0.$$

Denote

$$F_n(t) = \int_{D_{\lambda_n^t}(x_n)} (\sigma_n + |\nabla_{\widetilde{g}_n}\phi_{\alpha_n}|^2)^{\alpha_n - 1}|\nabla\phi_{\alpha_n}|^2 dx,$$

$$F_{r,n}(t) = \int_{D_{\lambda_n^t}(x_n)\setminus D_{\lambda_n^{t_0}}(x_n)} (\sigma_n + |\nabla_{\widetilde{g}_n}\phi_{\alpha_n}|^2)^{\alpha_n - 1}|\frac{\partial\phi_{\alpha_n}}{\partial r}|^2 dx$$

and

$$F_{\theta,n}(t)=\int_{D_{\lambda_n^t}(x_n)\setminus D_{\lambda_n^{t_0}}(x_n)}(\sigma_n+|\nabla_{\widetilde{g}_n}\phi_{\alpha_n}|^2)^{\alpha_n-1}|x-x_n|^{-2}|\frac{\partial\phi_{\alpha_n}}{\partial\theta}|^2dx,$$
$$0<t\le t_0\le 1.$$

By Corollary 3.6, for $t\in[\epsilon,t_0]$, we have

$$(1-\frac{1}{2\alpha_n})\frac{d}{dt}F_{r,n}(t)-\frac{1}{2\alpha_n}\frac{d}{dt}F_{\theta,n}=\frac{\alpha_n-1}{\alpha_n}\log\lambda_nF_n(t)+O(\lambda_n^t\log\lambda_n).$$

Then

$$(1-\frac{1}{2\alpha_n})F_{r,n}(t)-\frac{1}{2\alpha_n}F_{\theta,n}=\frac{1}{2}\int_{t_0}^t\left[-\frac{1}{\alpha_n}\log\lambda_n^{2(1-\alpha_n)}F_n(t)+O(\lambda_n^t\log\lambda_n)\right]dt. \tag{3.22}$$

It is easy to check that

$$\frac{1}{2}\int_{t_0}^t\left[-\frac{1}{\alpha_n}\log\lambda_n^{2(1-\alpha_n)}F_n(t)+O(\lambda_n^t\log\lambda_n)\right]dt$$

is compact in $C^0([\epsilon,t_0])$. Combining these with the fact that $F_{\theta,n}(t)\to 0$ by Corollary 3.10, we know that the sequences $(1-\frac{1}{2\alpha_n})F_{r,n}(t)-\frac{1}{2\alpha_n}F_{\theta,n}$, $\{F_n(t)\}$, $\{F_{r,n}(t)\}$ and $\{F_{\theta,n}(t)\}$ are compact in the $C^0([\epsilon,t_0])$ topology for any $\epsilon>0$. Thus, there exist two functions F_∞ and $F_{r,\infty}$ which belong to $C^0([\epsilon,t_0])$ such that

$$F_n\to F_\infty \quad \textit{and} \quad F_{r,n}\to F_{r,\infty} \quad \textit{in } C^0([\epsilon,t_0])$$

as $n\to\infty$.

LEMMA 3.11. *The functionals $F_\infty(t)$ and $F_{r,\infty}(t)$ satisfy the following relation:*

$$F_{r,\infty}(t)=\mu^{t_0-t}F_\infty(t_0)-F_\infty(t_0)\quad \textit{for any } 0<t\le t_0<1. \tag{3.23}$$

Moreover, we have:

$$\lim_{t_0\to 1^-}F_\infty(t_0)=\mu l_\infty E(w^1)\quad \textit{for any } 0<t<1, \tag{3.24}$$

where $\mu:=\lim_{n\to\infty}(\lambda_n)^{2-2\alpha_n}$, $l_\infty:=\lim_{n\to\infty}(\widetilde{l}_n)^{2-2\alpha_n}$ and w^1 is the first bubble in (3.19).

PROOF. Letting $n\to\infty$ in (3.22), we get

$$F_{r,\infty}(t)=-\log\mu\int_{t_0}^tF_\infty(s)ds \textit{ for any } 0<t<t_0. \tag{3.25}$$

Since

$$F_n(t)=F_{r,n}(t)+F_{\theta,n}(t)+F_n(t_0),$$

letting $n\to\infty$, Corollary 3.10 yields

$$F_\infty(t)=F_{r,\infty}(t)+F_\infty(t_0).$$

Then,

$$F_{r,\infty}(t)=-\log\mu\int_{t_0}^tF_\infty(s)ds=-\log\mu\int_{t_0}^t(F_{r,\infty}(s)+F_\infty(t_0))\,ds,$$

which implies $F_{r,\infty}(t)\in C^1$ and

$$\frac{d}{dt}F_{r,\infty}(t)=-\log\mu(F_{r,\infty}(t)+F_\infty(t_0)).$$

Thus,

$$F_{r,\infty}(t) = \mu^{t_0 - t} F_\infty(t_0) - F_\infty(t_0).$$

For equation (3.24), by Corollary 3.6, we have

$$\begin{aligned}
&\left| F_n(t_0) - \int_{D_{\lambda_n R}(x_n)} (\sigma_n + |\nabla_{\widetilde{g}_n}\phi_{\alpha_n}|^2)^{\alpha_n - 1} |\nabla \phi_{\alpha_n}|^2 dx \right| \\
&\leq C \int_{D_{\lambda_n^{t_0}}(x_n) \setminus D_{\lambda_n R}(x_n)} (\sigma_n + |\nabla_{\widetilde{g}_n}\phi_{\alpha_n}|^2)^{\alpha_n - 1} |x - x_n|^{-2} |\frac{\partial \phi_{\alpha_n}}{\partial \theta}|^2 dx \\
&\quad + C \int_{\lambda_n R}^{\lambda_n^{t_0}} \frac{\alpha_n - 1}{r} dr + C(\lambda_n^{t_0} - \lambda_n R) \\
&\leq C \int_{D_{\lambda_n^{t_0}}(x_n) \setminus D_{\lambda_n R}(x_n)} (\sigma_n + |\nabla_{\widetilde{g}_n}\phi_{\alpha_n}|^2)^{\alpha_n - 1} |x - x_n|^{-2} |\frac{\partial \phi_{\alpha_n}}{\partial \theta}|^2 dx \\
&\quad + C\left[(t_0 - 1)(\alpha_n - 1)\log \lambda_n - (\alpha_n - 1)\log R\right] + C(\lambda_n^{t_0} - \lambda_n R).
\end{aligned}$$

Combining this with Corollary 3.10, we get

$$\begin{aligned}
\lim_{t_0 \to 1^-} F_\infty(t_0) &= \lim_{R \to \infty} \lim_{n \to \infty} \int_{D_{\lambda_n R}(x_n)} (\sigma_n + |\nabla_{\widetilde{g}_n}\phi_{\alpha_n}|^2)^{\alpha_n - 1} |\nabla \phi_{\alpha_n}|^2 dx \\
&= \lim_{R \to \infty} \lim_{n \to \infty} \int_{D_R(0)} (\sigma_n \lambda_n^2 (\widetilde{l}_n)^2 + \frac{\widetilde{l}_n^2}{\widetilde{\rho}_n^2} |\nabla u_n|^2)^{\alpha_n - 1} \\
&\quad \times (\lambda_n \widetilde{l}_n)^{2 - 2\alpha_n} |\nabla u_n|^2 dx \\
&= \mu l_\infty \lim_{R \to \infty} \int_{D_R(0)} |\nabla w^1|^2 dx \\
&= \mu l_\infty E(w^1(x)).
\end{aligned}$$

□

LEMMA 3.12. *Under the assumptions of Lemma 3.8, for any $t \in (0,1)$, there holds*

$$\lim_{n \to \infty} \int_{D_{\lambda_n^t}(x_n)} (\sigma_n + |\nabla_{\widetilde{g}_n}\phi_{\alpha_n}|^2)^{\alpha_n - 1} |\nabla \phi_{\alpha_n}|^2 dx = \mu^{2-t} l_\infty E(w^1).$$

PROOF. By Lemma 3.11 and Corollary 3.10, it is easy to see that

$$\begin{aligned}
&\lim_{n \to \infty} \int_{D_{\lambda_n^t}(x_n)} (\sigma_n + |\nabla_{\widetilde{g}_n}\phi_{\alpha_n}|^2)^{\alpha_n - 1} |\nabla \phi_{\alpha_n}|^2 dx = F_{r,\infty}(t) + F_\infty(t_0) \\
&= \mu^{t_0 - t} F_\infty(t_0).
\end{aligned}$$

Then the conclusion of the lemma follows from letting $t_0 \nearrow 1$ and Lemma 3.11. □

In the end of this section, we complete the proof of Theorem 3.1.

Proof of Theorem 3.1. By Corollary 3.6 and Corollary 3.10, we have

$$\begin{aligned}&\int_{D_\delta(x_n)\setminus D_{\lambda_n^t}(x_n)}(\sigma_n+|\nabla_{\widetilde{g}_n}\phi_{\alpha_n}|^2)^{\alpha_n-1}|\nabla\phi_{\alpha_n}|^2dx\\&\le C\int_{D_\delta(x_n)\setminus D_{\lambda_n^t}(x_n)}(\sigma_n+|\nabla_{\widetilde{g}_n}\phi_{\alpha_n}|^2)^{\alpha_n-1}|x-x_n|^{-2}|\frac{\partial\phi_{\alpha_n}}{\partial\theta}|^2dx\\&\quad+C\int_{\lambda_n^t}^{\delta}\frac{\alpha_n-1}{r}dr+C(\delta-\lambda_n^t)\\&\le C\int_{D_\delta(x_n)\setminus D_{\lambda_n^t}(x_n)}(\sigma_n+|\nabla_{\widetilde{g}_n}\phi_{\alpha_n}|^2)^{\alpha_n-1}|x-x_n|^{-2}|\frac{\partial\phi_{\alpha_n}}{\partial\theta}|^2dx\\&\quad+C\left[(\alpha_n-1)\log\delta-t(\alpha_n-1)\log\lambda_n\right]+C(\delta-\lambda_n^t),\end{aligned}$$

which implies

$$\lim_{\delta\to0}\lim_{t\to0}\lim_{n\to\infty}\int_{D_\delta(x_n)\setminus D_{\lambda_n^t}(x_n)}(\sigma_n+|\nabla_{\widetilde{g}_n}\phi_{\alpha_n}|^2)^{\alpha_n-1}|\nabla\phi_{\alpha_n}|^2dx=0.$$

By Lemma 3.12, we get

$$\begin{aligned}&\lim_{\delta\to0}\lim_{n\to\infty}\int_{D_\delta(x_n)}(\sigma_n+|\nabla_{\widetilde{g}_n}\phi_{\alpha_n}|^2)^{\alpha_n-1}|\nabla\phi_{\alpha_n}|^2dx\\&=\lim_{\delta\to0}\lim_{t\to0}\lim_{n\to\infty}\left(\int_{D_\delta(x_n)\setminus D_{\lambda_n^t}(x_n)}(\sigma_n+|\nabla_{\widetilde{g}_n}\phi_{\alpha_n}|^2)^{\alpha_n-1}|\nabla\phi_{\alpha_n}|^2dx\right.\\&\quad\left.+\int_{D_{\lambda_n^t}(x_n)}(\sigma_n+|\nabla_{\widetilde{g}_n}\phi_{\alpha_n}|^2)^{\alpha_n-1}|\nabla\phi_{\alpha_n}|^2dx\right)\\&=\mu^2l_\infty E(w^1).\end{aligned}$$

We finished the proof of Theorem 3.1. □

3.4. Refined neck analysis on the second type neck

In this subsection, we will show that the second type neck also converges to a geodesic in the target manifold. Furthermore, we will give a length formula in the case that there is only one bubble.

Before stating the next lemma, we give some notations. Denote

$$P_i':=D_{e^{(i+1)L}\lambda_nR}(x_n)\setminus D_{e^{iL}\lambda_nR}(x_n).$$

Without loss of generality, we assume

$$D_\delta(x_n)\setminus D_{\lambda_nR}(x_n)=\cup_{i=0}^{k_n'}P_i'$$

for some positive integer k_n' which goes to infinity as $n\to\infty$. Similarly to Lemma 2.11, we have the following lemma.

LEMMA 3.13. *Under the assumptions of Theorem 3.2, we have*

$$\left(\int_{P_i'}|x-x_n|^{-2}|\frac{\partial\phi_{\alpha_n}}{\partial\theta}|^2dx\right)^{\frac12}\le\left((\alpha_n-1)+e^{-\frac{iL}{2}}+e^{-\frac{(k_n'-i)L}{2}}\right)o(n,\delta,R),$$

where $\lim_{\delta\to0}\lim_{R\to\infty}\lim_{n\to\infty}o(n,\delta,R)=0$.

PROOF. With the help of Lemma 3.4, Lemma 3.7 and (3.7), (3.8), (3.9), the proof is almost the same as Lemma 2.11.

Recalling

$$\phi^*_{\alpha_n}(r) = \frac{1}{2\pi}\int_0^{2\pi} \phi_{\alpha_n}(r,\theta)d\theta, \tag{3.26}$$

by a similar computation as in the proof of Lemma 2.11, we can get

$$\begin{aligned}\Delta(\phi_{\alpha_n} - \phi^*_{\alpha_n}) &= A^1(\phi_{\alpha_n} - \phi^*_{\alpha_n}) + A^2\nabla(\phi_{\alpha_n} - \phi^*_{\alpha_n}) \\ &\quad + \frac{1}{2\pi}\int_0^{2\pi} A^3(\phi_{\alpha_n} - \phi^*_{\alpha_n}) + A^4\nabla(\phi_{\alpha_n} - \phi^*_{\alpha_n})d\theta \\ &\quad - (\alpha_n - 1)\frac{\nabla|\nabla_{\widetilde{g}_n}\phi_{\alpha_n}|^2\nabla\phi_{\alpha_n}}{1+|\nabla_{\widetilde{g}_n}\phi_{\alpha_n}|^2} \\ &\quad + \frac{1}{2\pi}\int_0^{2\pi}(\alpha_n - 1)\frac{\nabla|\nabla_{\widetilde{g}_n}\phi_{\alpha_n}|^2\nabla\phi_{\alpha_n}}{1+|\nabla_{\widetilde{g}_n}\phi_{\alpha_n}|^2}d\theta,\end{aligned} \tag{3.27}$$

where

$$|A^1| + |A^3| \le C(N)|\nabla\phi_{\alpha_n}|^2, \quad |A^2| + |A^4| \le C(N)|\nabla\phi_{\alpha_n}|.$$

By (3.20) and Lemma 3.4, we have that for any $\epsilon > 0$, there holds

$$\|\ |x|(|A^2| + |A^4|) + |x|^2(|A^1| + |A^3|)\ \|_{L^\infty(D_\delta(x_n)\setminus D_{\lambda_n R}(x_n))} \le \epsilon,$$

when n is big enough.

Substituting $u = u_n := \phi_{\alpha_n} - \phi^*_{\alpha_n}$ and

$$f = -(\alpha_n - 1)\frac{\nabla|\nabla_{\widetilde{g}_n}\phi_{\alpha_n}|^2\nabla\phi_{\alpha_n}}{1+|\nabla_{\widetilde{g}_n}\phi_{\alpha_n}|^2} + \frac{1}{2\pi}\int_0^{2\pi}(\alpha_n - 1)\frac{\nabla|\nabla_{\widetilde{g}_n}\phi_{\alpha_n}|^2\nabla\phi_{\alpha_n}}{1+|\nabla_{\widetilde{g}_n}\phi_{\alpha_n}|^2}d\theta$$

in Lemma 3.7 and noting that

$$\begin{aligned}\||x||f|\|_{L^2(P'_i)} &\le C(\alpha_n - 1)\left(\||x||\nabla^2\phi_{\alpha_n}|\|_{L^2(P'_i)} + \|\nabla\phi_{\alpha_n}\|_{L^2(P'_i)}\right) \\ &\le C(\alpha_n - 1)\|\nabla\phi_{\alpha_n}\|_{L^2(P'_{i-1}\cup P'_i\cup P'_{i+1})},\ i = 1,\dots,k'_n,\end{aligned}$$

where we used (3.8), (3.9) and Lemma 3.4, we obtain the energy decay in the θ-direction,

$$\begin{aligned}\|\frac{\partial\phi_{\alpha_n}}{\partial\theta}\|_{L^2(P'_i)} &\le \|\nabla(\phi_{\alpha_n} - \phi^*_{\alpha_n})\|_{L^2(P'_i)} \\ &\le C(\|\phi_{\alpha_n} - \phi^*_{\alpha_n}\|_{L^2(P'_{i-1}\cup P'_i\cup P'_{i+1})} \\ &\quad + \|\Delta(\phi_{\alpha_n} - \phi^*_{\alpha_n})\|_{L^2(P'_{i-1}\cup P'_i\cup P'_{i+1})}) \\ &\le C(\alpha_n - 1)\max_{j=1,\dots,k'_n}\|\nabla\phi_{\alpha_n}\|_{L^2(P'_j)} \\ &\quad + C\big(\|u_n\|_{L^2(P'_1)} + \|u_n\|_{L^2(P_{k'_n})}\big)\big(e^{-\frac{1}{2}(k'_n - i)L} + e^{-\frac{1}{2}iL}\big) \\ &\le C\left((\alpha_n - 1) + e^{-\frac{1}{2}(k'_n - i)L} + e^{-\frac{1}{2}iL}\right)o(n),\end{aligned} \tag{3.28}$$

where we used the fact that

$$\|u_n\|_{L^2(P_1)} + \|u_n\|_{L^2(P_{k'_n})} = o(n).$$

□

LEMMA 3.14. *Under the assumption of Theorem 3.2, suppose that $\nu > 1$ and t_n is a positive number such that $0 < t_1 \le t_n \le t_2 < 1$. Then we have*

$$\frac{1}{\sqrt{\alpha_n - 1}}\left(\phi_{\alpha_n}(x_n + \lambda_n^{t_n} x) - \phi_{\alpha_n}(x_n + (\lambda_n^{t_n}, 0))\right) \to \overrightarrow{a} \log|x|$$

strongly in $C^j(D_K(0) \setminus D_{\frac{1}{K}}(0), \mathbb{R}^N)$ for any $K > 0$ and integer j, where $\overrightarrow{a} \in T_y N$ is a vector in $\mathbb{R}^N$ with

$$|\overrightarrow{a}| = \mu^{1-\lim_{n\to\infty} t_n}\sqrt{\frac{E(w^1)}{\pi}}.$$

PROOF. Set

$$u_n(x) = \phi_{\alpha_n}(x_n + \lambda_n^{t_n} x)$$

and

$$v_n(x) = \frac{1}{\sqrt{\alpha_n - 1}}\left(\phi_{\alpha_n}(x_n + \lambda_n^{t_n} x) - \phi_{\alpha_n}(x_n + (\lambda_n^{t_n}, 0))\right).$$

By Corollary 3.6, we get

$$\begin{aligned}
&\int_{D_{2^k\lambda_n^{t_n}}(x_n)\setminus D_{2^{-k}\lambda_n^{t_n}}(x_n)} |\nabla\phi_{\alpha_n}|^2 dx \\
&\le C\int_{D_{2^k\lambda_n^{t_n}}(x_n)\setminus D_{2^{-k}\lambda_n^{t_n}}(x_n)} (\sigma_n + |\nabla_{\widetilde{g}_n}\phi_{\alpha_n}|^2)^{\alpha_n - 1}|\nabla\phi_{\alpha_n}|^2 dx \\
&\le C\int_{D_{2^k\lambda_n^{t_n}}(x_n)\setminus D_{2^{-k}\lambda_n^{t_n}}(x_n)} (\sigma_n + |\nabla_{\widetilde{g}_n}\phi_{\alpha_n}|^2)^{\alpha_n - 1}|x - x_n|^{-2}\left|\frac{\partial\phi_{\alpha_n}}{\partial\theta}\right|^2 dx \\
&\quad + C\int_{2^{-k}\lambda_n^{t_n}}^{2^k\lambda_n^{t_n}} \frac{\alpha_n - 1}{r} dr + C(k)\lambda_n^{t_n} \\
&\le C\left\| |x - x_n|^{-1}\frac{\partial\phi_{\alpha_n}}{\partial\theta}\right\|^2_{L^2(D_{2^k\lambda_n^{t_n}}(x_n)\setminus D_{2^{-k}\lambda_n^{t_n}}(x_n))} + C(k)(\alpha_n - 1 + \lambda_n^{t_n}).
\end{aligned} \tag{3.29}$$

Without loss of generality, we assume $\delta = e^{k_n' L}\lambda_n R$ for some integer k_n' which goes to infinity as $n \to \infty$. Let $i_1, i_2 \in [0, k_n']$ be two integers such that

$$e^{i_1 L}\lambda_n R \le 2^{-k}\lambda_n^{t_n} < e^{(i_1+1)L}\lambda_n R \quad \text{and} \quad e^{i_2 L}\lambda_n R \le 2^k\lambda_n^{t_n} < e^{(i_2+1)L}\lambda_n R.$$

Then we have

$$e^{-i_1 L} \le 2^k R e^L \lambda_n^{1-t_n}, \quad e^{-(k_n' - i_2)L} \le \frac{2^k}{\delta}\lambda_n^{t_n}, \quad i_2 - i_1 \le \frac{2k\log 2}{L} + 1.$$

According to Lemma 3.13, there holds

$$\begin{aligned}
&\left\| |x - x_n|^{-1}\frac{\partial\phi_{\alpha_n}}{\partial\theta}\right\|_{L^2(D_{2^k\lambda_n^{t_n}}(x_n)\setminus D_{2^{-k}\lambda_n^{t_n}}(x_n))} \\
&\le \sum_{i=i_1}^{i_2}\left\| |x - x_n|^{-1}\frac{\partial\phi_{\alpha_n}}{\partial\theta}\right\|_{P_i'} \\
&\le \left((\alpha_n - 1)(i_2 - i_1) + e^{-\frac{i_1 L}{2}} + e^{-\frac{(k_n' - i_2)L}{2}}\right) o(n, \delta, R) \\
&\le \left((\alpha_n - 1) + \sqrt{2^k R e^L}\lambda_n^{\frac{1-t_n}{2}} + \sqrt{\frac{2^k}{\delta}}\lambda_n^{\frac{t_n}{2}}\right) o(n, \delta, R) \le (\alpha_n - 1) o(n, \delta, R),
\end{aligned}$$

where the last inequality follows from the fact that $\nu > 1$ which implies

$$\lambda_n = o((\alpha_n - 1)^m) \quad \textit{for any } m > 0.$$

Then (3.29) yields

$$\int_{D_{2^k\lambda_n^{t_n}}(x_n)\setminus D_{2^{-k}\lambda_n^{t_n}}(x_n)} |\nabla\phi_{\alpha_n}|^2 dx \le C(\alpha_n - 1).$$

By Lemma 3.4, we get

$$\|\nabla u_n\|_{C^0(D_{2^k}(0)\setminus D_{2^{-k}}(0))} + \|\nabla^2 u_n\|_{C^0(D_{2^k}(0)\setminus D_{2^{-k}}(0))} \le C(k)\sqrt{\alpha_n - 1}, \tag{3.30}$$

which implies

$$\|\nabla v_n\|_{C^0(D_{2^k}(0)\setminus D_{2^{-k}}(0))} + \|\nabla^2 v_n\|_{L^\infty(D_{2^k}(0)\setminus D_{2^{-k}}(0))} \le C(k).$$

Noting that $v_n(1,0) = 0$, thus

$$\|v_n\|_{C^2(D_{2^k}(0)\setminus D_{2^{-k}}(0))} \le C(k).$$

Since v_n satisfies the following equation

$$\Delta v_n + \sqrt{\alpha_n - 1}O(|\nabla v_n|^2) + (\alpha_n - 1)\left(O(|\nabla^2 v_n|) + O(|\nabla v_n|)\right) = 0,$$

then there exists a subsequence of v_n (without changing the notation) such that

$$v_n \to v_0 \quad in \quad C^1_{loc}(\mathbb{R}^2 \setminus \{0\}) \tag{3.31}$$

where v_0 satisfies

$$\Delta v_0 = 0.$$

Moreover, Lemma 3.13 tells us that $v_0(x) = v_0(|x|)$. Set

$$v_0 = \overrightarrow{a}\log r = (a_1, \ldots, a_N)\log r.$$

By Corollary 3.6 and Lemma 3.13, we have

$$\begin{aligned}
&\frac{1}{\alpha_n - 1}\int_{D_{2\lambda_n^{t_n}}(x_n)\setminus D_{\lambda_n^{t_n}}(x_n)} (\sigma_n + |\nabla_{\widetilde{g}_n}\phi_{\alpha_n}|^2)^{\alpha_n - 1}|\nabla\phi_{\alpha_n}|^2 dx \\
&= \frac{2}{2\alpha_n - 1}\int_{\lambda_n^{t_n}}^{2\lambda_n^{t_n}} \frac{1}{t}F_n(\log_{\lambda_n} t)dt + o(1) \to 2\log 2 F_\infty(\lim_{n\to\infty} t_n),
\end{aligned}$$

where we used the fact that

$$\lambda_n^{t_n} = o((\alpha - 1)^m)$$

for any $m > 0$, since $\nu > 1$.

On the other hand, there also holds

$$\begin{aligned}
&\frac{1}{\alpha_n - 1}\int_{D_{2\lambda_n^{t_n}}(x_n)\setminus D_{\lambda_n^{t_n}}(x_n)} (\sigma_n + |\nabla_{\widetilde{g}_n}\phi_{\alpha_n}|^2)^{\alpha_n - 1}|\nabla\phi_{\alpha_n}|^2 dx \\
&= \int_{D_2(0)\setminus D_1(0)} (\sigma_n + |\nabla_{\widetilde{g}_n} v_n|^2 \frac{\alpha_n - 1}{\lambda_n^{2t_n}})^{\alpha_n - 1}|\nabla v_n|^2 dx \\
&\to 2\pi\log 2|\overrightarrow{a}|^2 l_\infty \mu^{\lim_{n\to\infty} t_n}.
\end{aligned}$$

Therefore,

$$|\overrightarrow{a}|^2 = \frac{1}{\pi l_\infty}\mu^{-\lim_{n\to\infty} t_n}F_\infty(\lim_{n\to\infty} t_n) = \frac{E(w^1)}{\pi}\mu^{2 - 2\lim_{n\to\infty} t_n},$$

where the last equality follows from Lemma 3.12 and we proved the lemma. □

LEMMA 3.15. *Under the assumptions of Theorem 3.2, suppose that $\nu > 1$. Then we have*

(1)

$$\left\| \frac{1}{\sqrt{\alpha_n - 1}} \left(\phi_{\alpha_n} - \frac{1}{2\pi} \int_0^{2\pi} \phi_{\alpha_n} d\theta \right) \right\|_{C^0(D_{\lambda_n^{t_1}}(x_n) \setminus D_{\lambda_n^{t_2}}(x_n))} + \left\| \frac{|x - x_n|}{\sqrt{\alpha_n - 1}} \nabla \left(\phi_{\alpha_n} - \frac{1}{2\pi} \int_0^{2\pi} \phi_{\alpha_n} d\theta \right) \right\|_{C^0(D_{\lambda_n^{t_1}}(x_n) \setminus D_{\lambda_n^{t_2}}(x_n))} \to 0;$$

(2)

$$\int_{\lambda_n^t}^{2\lambda_n^t} \frac{1}{\sqrt{\alpha_n - 1}} |\frac{\partial \phi_{\alpha_n}}{\partial r}| dr \to \log 2 \mu^{1-t} \sqrt{\frac{E(w^1)}{\pi}} \quad \text{in } C^0([t_1, t_2])$$

and

$$\frac{1}{\sqrt{\alpha_n - 1}} (r |\frac{\partial \phi_{\alpha_n}}{\partial r}|)(\lambda_n^t, \theta) \to \mu^{1-t} \sqrt{\frac{E(w^1)}{\pi}} \quad \text{in } C^0([t_1, t_2]),$$

$$\frac{1}{\sqrt{\alpha_n - 1}} (r^{-1} |\frac{\partial \phi_{\alpha_n}}{\partial \theta}|)(\lambda_n^t, \theta) \to 0 \quad \text{in } C^0([t_1, t_2]).$$

PROOF. For statement (1), if it is false, then there would exist $t_n \in [t_1, t_2]$ and $\theta_n \in [0, 2\pi]$ such that

(3.32)
$$\frac{1}{\sqrt{\alpha_n - 1}} \left(\lambda_n^{t_n} |\nabla \phi_{\alpha_n}(\lambda_n^{t_n}, \theta_n) - \nabla \phi^*_{\alpha_n}(\lambda_n^{t_n})| + |\phi_{\alpha_n}(\lambda_n^{t_n}, \theta_n) - \phi^*_{\alpha_n}(\lambda_n^{t_n})| \right) \geq b > 0,$$

where $\phi^*_{\alpha_n} := \frac{1}{2\pi} \int_0^{2\pi} \phi_{\alpha_n} d\theta$.

Let $\lim_{n\to\infty} \theta_n = \theta_0$, then it is easy to see that

$$|\nabla v_0(1, \theta_0) - \nabla v_0^*(1)| + |v_0(1, \theta_0) - v_0^*(1)| = 0,$$

where v_0 is the limit of v_n as in (3.31) and v_0^* is the integral average of v_0, similarly to the definition of ϕ^*_α.

However, by the fact that $v_n \to v_0$ in $C^1_{loc}(\mathbb{R}^2 \setminus \{0\})$, we have

$$\begin{aligned} b &\leq \lim_{n\to\infty} \frac{1}{\sqrt{\alpha_n - 1}} \left(\lambda_n^{t_n} |\nabla \phi_{\alpha_n}(\lambda_n^{t_n}, \theta_n) - \nabla \phi^*_{\alpha_n}(\lambda_n^{t_n})| + |\phi_{\alpha_n}(\lambda_n^{t_n}, \theta_n) - \phi^*_{\alpha_n}(\lambda_n^{t_n})| \right) \\ &= \lim_{n\to\infty} |\nabla v_n(1, \theta_n) - \nabla v_n^*(1)| + \lim_{n\to\infty} |v_n(1, \theta_n) - v_n^*(1)| \\ &= |\nabla v_0(1, \theta_0) - \nabla v_0^*(1)| + |v_0(1, \theta_0) - v_0^*(1)| = 0, \end{aligned}$$

which is a contradiction. So, the statement (1) holds.

For statement (2), we just need to prove the second claim, since the proof of the other two claims are similar. In fact, if it is false, then there exist $t_n \in [t_1, t_2]$ and $\theta_n \in [0, 2\pi]$ such that

$$\left| \frac{1}{\sqrt{\alpha_n - 1}} (r |\frac{\partial \phi_{\alpha_n}}{\partial r}|)(\lambda_n^{t_n}, \theta_n) - \mu^{1-t_n} \sqrt{\frac{E(w^1)}{\pi}} \right| \geq b > 0. \tag{3.33}$$

However, by Lemma 3.14, we have

$$\frac{\lambda_n^{t_n}}{\sqrt{\alpha_n - 1}} \frac{\partial \phi_{\alpha_n}}{\partial r}(\lambda_n^{t_n}, \theta_n) \to \overrightarrow{a},$$

where $|\overrightarrow{a}| = \mu^{1-\lim_{n\to\infty} t_n}\sqrt{\frac{E(w^1)}{\pi}}$. This yields the following

$$\frac{1}{\sqrt{\alpha_n - 1}}(r|\frac{\partial \phi_{\alpha_n}}{\partial r}|)(\lambda_n^{t_n}, \theta_n) \to \mu^{1-\lim_{n\to\infty} t_n}\sqrt{\frac{E(w^1)}{\pi}},$$

which contradicts to (3.33). We finished the proof of this lemma. □

For $0 < t_1 < t_2 < 1$, define the following curve:

$$\phi^*_{\alpha_n}(r) := \frac{1}{2\pi}\int_0^{2\pi} \phi_{\alpha_n}(r,\theta)d\theta : [\lambda_n^{t_2}, \lambda_n^{t_1}] \to \mathbb{R}^K, \tag{3.34}$$

which is denoted by γ_n, where (r, θ) is the polar coordinate around the point x_n.

Denote

$$\ddot{\phi}^*_{\alpha_n} := (\frac{d}{dr})^2\phi_{\alpha_n}, \quad \dot{\phi}^*_{\alpha_n} := \frac{d}{dr}\phi_{\alpha_n}.$$

By (3.7)-(3.9) and a direct computation, we have

$$\begin{aligned}
\ddot{\phi}^*_{\alpha_n} &= \frac{1}{2\pi}\int_0^{2\pi} \frac{\partial^2}{\partial r^2}\phi_{\alpha_n}(r,\theta)d\theta \\
&= \frac{1}{2\pi}\int_0^{2\pi} \Delta\phi_{\alpha_n}(r,\theta)d\theta - \frac{1}{2\pi}\int_0^{2\pi} \frac{1}{r}\frac{\partial}{\partial r}\phi_{\alpha_n}(r,\theta)d\theta \\
&= -\frac{1}{2\pi}\int_0^{2\pi} A(\phi_{\alpha_n})(d\phi_{\alpha_n}, d\phi_{\alpha_n})d\theta \\
&\quad + \frac{\alpha_n - 1}{2\pi}\int_0^{2\pi} O(|\nabla^2\phi_{\alpha_n}| + |\nabla\phi_{\alpha_n}|)d\theta - \frac{1}{r}\dot{\phi}^*_{\alpha_n}.
\end{aligned} \tag{3.35}$$

PROPOSITION 3.16. *Under the assumptions of Theorem 3.2, suppose that $\nu > 1$. Then after passing to a subsequence, the sequence of the curves $\{\gamma_n\}$ in $\mathbb{R}^N$, which is defined by $\phi^*_{\alpha_n}$, and parameterized by its arc length, converges to a curve $\gamma \subset N$ as $n \to \infty$, where γ is a geodesic on N, i.e.*

$$\frac{d^2\gamma}{ds^2} + A(\gamma)(\frac{d\gamma}{ds}, \frac{d\gamma}{ds}) = 0.$$

PROOF. Denote the induced second fundamental form of γ_n in $\mathbb{R}^K$ by A_{γ_n}. Set

$$\begin{aligned}
G_n :&= \ddot{\gamma}_n + \frac{\dot{\gamma}_n}{r} \\
&= \frac{1}{2\pi}\int_0^{2\pi} A(\phi_{\alpha_n})(d\phi_{\alpha_n}, d\phi_{\alpha_n})d\theta + \frac{\alpha_n - 1}{2\pi}\int_0^{2\pi} O(|\nabla^2\phi_{\alpha_n}| + |\nabla\phi_{\alpha_n}|)d\theta.
\end{aligned}$$

By Lemma 3.4, it is easy to see that γ_n converges to some curve on N denoted by γ. Next, we will show that γ is just a geodesic on N.

Let s be the arc length parameter of γ_α, i.e.

$$s(r) = \int_{\lambda_n^{t_n}}^r |\dot{\gamma}_n(\tau)|d\tau = \int_{\lambda_n^{t_n}}^r |\dot{\phi}^*_{\alpha_n}(\tau)|d\tau,$$

where $t_n \in [t_1, t_2]$.

By Lemma 3.15, for any $t_1 \in (0,1)$, we have

$$\int_{\lambda_n^{t_n}}^{\lambda_n^{t_1}} |\dot{\gamma}_n| dr \le \int_{\lambda_n^{t_n}}^{\lambda_n^{t_1}} \frac{\sqrt{\alpha_n - 1}}{r} \mu \sqrt{\frac{E(w^1)}{\pi}} dr$$

$$= (t_1 - t_n)\sqrt{\alpha_n - 1} \log \lambda_n \mu \sqrt{\frac{E(w^1)}{\pi}} \to (\lim_{n\to\infty} t_n - t_1) \log \nu \mu \sqrt{\frac{E(w^1)}{\pi}}$$

as $n \to 1$. So, if $\nu = \infty$, then for any $s \in (0, \infty)$,

$$\gamma_n\big|_{[0,s]} \subset \gamma_n\big|_{[\lambda_n^{t_2}, \lambda_n^{t_1}]}.$$

If $\nu \in (1, \infty)$, then $\mu = 1$ and noting that

$$\int_{\lambda_n^{t_n}}^{\lambda_n^{t_1}} |\dot{\gamma}_n| dr \ge \int_{\lambda_n^{t_n}}^{\lambda_n^{t_1}} \frac{\sqrt{\alpha_n - 1}}{r} \sqrt{\frac{E(w^1)}{\pi}} dr$$

$$= (t_1 - t_n)\sqrt{\alpha_n - 1} \log \lambda_n \sqrt{\frac{E(w^1)}{\pi}} \to (\lim_{n\to\infty} t_n - t_1) \log \nu \sqrt{\frac{E(w^1)}{\pi}}$$

as $n \to \infty$, there exists $t_1' \in (0, t_1)$ such that

$$\gamma_n\big|_{[0,s]} \subset \gamma_n\big|_{[\lambda_n^{t_2}, \lambda_n^{t_1'}]}$$

whenever $s \in \left(0, (\lim_{n\to\infty} t_n - t_1) \log \nu \sqrt{\frac{E(w^1)}{\pi}}\right)$.

Computing directly, we obtain

$$\begin{aligned}
-\frac{d^2\gamma_n}{ds^2} &= A_{\gamma_n}(\gamma_n(s))(\frac{d\gamma_n}{ds}, \frac{d\gamma_n}{ds}) = \frac{1}{|\dot{\gamma}_n|^2} A_{\gamma_n}(\gamma_n(r))(\frac{d\gamma_n}{dr}, \frac{d\gamma_n}{dr}) \\
&= -\frac{1}{|\dot{\gamma}_n|^2}(\ddot{\gamma}_n - \frac{\langle \ddot{\gamma}_n, \dot{\gamma}_n\rangle}{|\dot{\gamma}_n|^2}\dot{\gamma}_n) = -\frac{1}{|\dot{\gamma}_n|^2}(G_n - \frac{\langle G_n, \dot{\gamma}_n\rangle}{|\dot{\gamma}_n|^2}\dot{\gamma}_n) \\
&= \frac{1}{|\dot{\phi}^*_{\alpha_n}|^2} \frac{1}{2\pi} \int_0^{2\pi} A(\phi_{\alpha_n})(\nabla\phi_{\alpha_n}, \nabla\phi_{\alpha_n}) d\theta \\
&\quad + \frac{1}{|\dot{\phi}^*_{\alpha_n}|^2} \frac{1}{2\pi} \left(\int_0^{2\pi} (\alpha_n - 1) O(|\nabla^2\phi_{\alpha_n}| + |\nabla\phi_{\alpha_n}|) d\theta \right) \\
&\quad - \frac{\dot{\phi}^*_{\alpha_n}}{|\dot{\phi}_{\alpha_n}|^4} \frac{1}{2\pi} \int_0^{2\pi} \langle A(\phi_{\alpha_n})(\nabla\phi_{\alpha_n}, \nabla\phi_{\alpha_n}), \dot{\phi}^*_{\alpha_n}\rangle d\theta \\
&\quad - \frac{\dot{\phi}_{\alpha_n}}{|\dot{\phi}_{\alpha_n}|^4} \frac{1}{2\pi} \int_0^{2\pi} \langle (\alpha_n - 1) O(|\nabla^2\phi_{\alpha_n}| + |\nabla\phi_{\alpha_n}|), \dot{\phi}^*_{\alpha_n}\rangle d\theta \\
&:= \mathbf{I} + \mathbf{II} + \mathbf{III} + \mathbf{IV}. \qquad (3.36)
\end{aligned}$$

On one hand, by Lemma 3.15 and using the fact that

$$\langle A(\phi_{\alpha_n})(\nabla\phi_{\alpha_n}, \nabla\phi_{\alpha_n}), \frac{\partial\phi_{\alpha_n}}{\partial r}\rangle = 0,$$

we have

$$\begin{aligned}
\mathbf{III} &= \frac{\dot{\phi}^*_{\alpha_n}}{|\dot{\phi}^*_{\alpha_n}|^4} \frac{1}{2\pi} \int_0^{2\pi} \langle A(\phi_{\alpha_n})(\nabla\phi_{\alpha_n}, \nabla\phi_{\alpha_n}), \dot{\phi}_{\alpha_n} - \frac{\partial\phi_{\alpha_n}}{\partial r}\rangle d\theta \\
&\le \| \frac{|x - x_n|}{\sqrt{\alpha_n - 1}} \nabla(\phi_{\alpha_n} - \phi^*_{\alpha_n}) \|_{C^0(D_{\lambda_n^{t_1}}(x_n) \setminus D_{\lambda_n^{t_n}}(x_n))} \to 0 \qquad (3.37)
\end{aligned}$$

as $n \to \infty$.

On the other hand, by (3.30) and Lemma 3.15, we have

$$\mathbf{II}+\mathbf{IV} \le C\sqrt{\alpha_n - 1}\frac{|x-x_n|^2(|\nabla^2\phi_{\alpha_n}|+|\nabla\phi_{\alpha_n}|)}{\sqrt{\alpha_n-1}} \to 0 \tag{3.38}$$

as $n\to\infty$.

Therefore, we see that

$$|\frac{d^2\gamma_n}{ds^2}| \le C(N).$$

Thus, $\phi_{\alpha_n}(s)$ converges locally to a smooth vector valued function from $[0,s]$ into $\mathbb{R}^K$, denoted by $\gamma(s)$, in the sense of C^1, i.e., $\gamma_n|_{[\lambda_n^{t_2},\lambda_n^{t_1}]}$ converges locally to the curve γ.

Next, we will show that γ is a geodesic.

By Lemma 3.15, we obtain

$$\begin{aligned}
\mathbf{I} &= \frac{1}{|\dot\phi^*_{\alpha_n}|^2}\frac{1}{2\pi}\int_0^{2\pi} A(\phi_{\alpha_n})(\nabla\phi_{\alpha_n},\nabla\phi_{\alpha_n})d\theta\\
&= \frac{1}{|\dot\phi_{\alpha_n}|^2}\frac{1}{2\pi}\Big(\int_0^{2\pi}(A(\phi_{\alpha_n})-A(\phi^*_{\alpha_n}))(\nabla\phi_{\alpha_n},\nabla\phi_{\alpha_n})\\
&\qquad\qquad + A(\phi^*_{\alpha_n})(\nabla(\phi_{\alpha_n}-\phi^*_{\alpha_n}),\nabla\phi_{\alpha_n})d\theta\Big)\\
&\quad+\frac{1}{|\dot\phi^*_{\alpha_n}|^2}\frac{1}{2\pi}\int_0^{2\pi} A(\phi^*_{\alpha_n})(\nabla\phi^*_{\alpha_n},\nabla(\phi_{\alpha_n}-\phi^*_{\alpha_n}))d\theta\\
&\quad+\frac{1}{|\dot\phi^*_{\alpha_n}|^2}\frac{1}{2\pi}\int_0^{2\pi} A(\phi^*_{\alpha_n})(\nabla\phi^*_{\alpha_n},\nabla\phi^*_{\alpha_n})d\theta\\
&= \frac{1}{|\dot\phi^*_{\alpha_n}|^2}A(\phi^*_{\alpha_n})(\nabla\phi^*_{\alpha_n},\nabla\phi^*_{\alpha_n})\\
&\quad+O\left(\frac{|x-x_n||\nabla(\phi_{\alpha_n}-\phi^*_{\alpha_n})|}{\sqrt{\alpha_n-1}}\right)+O(|\phi_{\alpha_n}-\phi^*_{\alpha_n}|).
\end{aligned}\tag{3.39}$$

Combining this with (3.36) and Lemma 3.15, letting $n\to\infty$, we have

$$\frac{d\gamma}{ds}(s)-\frac{d\gamma}{ds}(0) = -\int_0^s A(\gamma)(\frac{d\gamma}{ds},\frac{d\gamma}{ds})$$

which implies

$$\frac{d^2\gamma}{ds^2} = -A(\gamma)(\frac{d\gamma}{ds},\frac{d\gamma}{ds}).$$

We finished the proof of this proposition. □

Now we prove Theorem 3.2.

Proof of Theorem 3.2. By Proposition 3.16, we just need to compute the length of the geodesic. Without loss of generality, we assume $\lambda_n^{t_1} = 2^{m_n}\lambda_n^{t_2}$ for some integer

$$m_n = \frac{t_1-t_2}{\log 2}\log\lambda_n,$$

which tends to infinity as $n\to\infty$. Then it is sufficient to consider the three cases listed in the statement of the theorem.

Case (3): $\nu=\infty$.

By Lemma 3.15, there holds

$$L(\gamma_n|_{D_{2^{k+1}\lambda_n^{t_2}}(x_n)\backslash D_{2^k\lambda_n^{t_2}}(x_n)}) \geq \sqrt{\alpha_n - 1}\left(\sqrt{\frac{E(w^1)}{\pi}}\log 2 + o(1)\right).$$

Then

$$L(\gamma_n) \geq Cm_n\sqrt{\alpha_n - 1} \geq -C\sqrt{\alpha_n - 1}\log\lambda_n \to \infty.$$

Case (2): $\nu \in (1, \infty)$.

Firstly, we prove the following two equalities:

$$\lim_{\delta\to 0}\lim_{t\to 0}\lim_{n\to\infty} Osc_{D_\delta(x_n)\backslash D_{\lambda_n^t}(x_n)}\ \phi_{\alpha_n} = 0 \tag{3.40}$$

and

$$\lim_{R\to\infty}\lim_{t\to 1}\lim_{n\to\infty} Osc_{D_{\lambda_n^t}(x_n)\backslash D_{\lambda_n R}(x_n)}\ \phi_{\alpha_n} = 0. \tag{3.41}$$

In fact, by Lemma 3.13, there holds

$$\left(\int_{P_i'}|x - x_\alpha|^{-2}\left|\frac{\partial\phi_\alpha}{\partial\theta}\right|^2 dx\right)^{\frac{1}{2}} \leq \left((\alpha_n - 1) + e^{-\frac{iL}{2}} + e^{-\frac{(k_n'-i)L}{2}}\right)o(n,\delta,R),$$

where $P_i' = D_{e^{(i+1)L}\lambda_n R}\setminus D_{e^{iL}\lambda_n R}$, $0 \leq i \leq k_n' := [\frac{\delta}{\lambda_n R}] - 1$.

Similarly to deriving (3.29), we have

$$\int_{P_i'}|\nabla\phi_{\alpha_n}|^2 dx \leq \left(e^{-iL} + e^{-(k_n'-i)L}\right)o(n,\delta,R) + C(\alpha_n - 1 + e^{iL}\lambda_n R). \tag{3.42}$$

According to Lemma 3.4, we get

$$Osc_{P_i'}\ \phi_{\alpha_n} \leq \left(e^{-\frac{iL}{2}} + e^{-\frac{(k_n'-i)L}{2}}\right)o(n,\delta,R) + C\left(\sqrt{\alpha_n - 1} + e^{\frac{iL}{2}}\sqrt{\lambda_n R}\right). \tag{3.43}$$

Let $i_t \in [1, k_n']$ be the positive integer such that

$$e^{i_t L}\lambda_n R \leq \lambda_n^t < e^{(i_t+1)L}\lambda_n R.$$

Then $k_n' - i_t \leq C\log\frac{\delta}{\lambda_n^t}$ and

$$Osc_{D_\delta(x_n)\backslash D_{\lambda_n^t}(x_n)}\ \phi_{\alpha_n} \leq \sum_{i=i_t}^{k_n'} Osc_{P_i'}\ \phi_{\alpha_n} \leq o(n,\delta,R) + C\left(\sqrt{\alpha_n - 1}\log\frac{\delta}{\lambda_n^t} + \sqrt{\delta}\right)$$

and (3.40) follows immediately. Similarly, we can prove (3.41).

Secondly, by Lemma 3.15, there holds

$$L(\gamma_n|_{D_{2^{k+1}\lambda_n^{t_2}}(x_n)\backslash D_{2^k\lambda_n^{t_2}}(x_n)}) = \sqrt{\alpha_n - 1}\left(\sqrt{\frac{E(w^1)}{\pi}}\log 2 + o(1)\right).$$

Then

$$\lim_{n\to\infty} L(\gamma_n|_{D_{\lambda_n^{t_1}}(x_n)\backslash D_{\lambda_n^{t_2}}(x_n)}) = \lim_{n\to\infty}\sqrt{\alpha_n - 1}\, m_n\left(\sqrt{\frac{E(w^1)}{\pi}}\log 2 + o(1)\right)$$

$$= (t_2 - t_1)\sqrt{\frac{E(w^1)}{\pi}}\log\nu,$$

which implies that the length of the limit curve is

$$L(\gamma) = \sqrt{\frac{E(w^1)}{\pi}}\log\nu.$$

Case (1): $\nu = 1$.
By (3.43), we have

$$Osc_{D_\delta(x_n)\setminus D_{\lambda_n R}(x_n)}\ \phi_{\alpha_n} \le \sum_{i=1}^{k'_n} Osc_{P'_i}\ \phi_{\alpha_n}$$
$$\le o(n,\delta,R) + C\left(\sqrt{\alpha_n - 1}\log\frac{\delta}{\lambda_n R} + \sqrt{\delta}\right)$$

which implies that

$$\lim_{\delta\to 0}\lim_{R\to\infty}\lim_{n\to\infty} Osc_{D_\delta(x_n)\setminus D_{\lambda_n R}(x_n)}\ \phi_{\alpha_n} = 0.$$

We finished the proof of Theorem 3.2. □

3.5. Compactness with bounded Morse index

In this subsection, we shll study the properties of the blow-up sequence with bounded Morse index. In particular, we will show that the energy identity holds if the target manifold satisfies some geometric and topological assumptions.

Let γ_n be the curve defined as in (3.34). Let s be the arc length parameter of γ_n such that $s(\lambda_n^{t_1}) = 0$, i.e.

$$s(r) = \int_{\lambda_n^{t_1}}^{r} |\dot{\gamma}_n(\tau)| d\tau = \int_{\lambda_n^{t_1}}^{r} |\dot{\phi}^*_{\alpha_n}(\tau)| d\tau.$$

If the limiting neck is a geodesic of infinite length, by Theorem 3.2, for any $a > 0$, after passing to a subsequence, the curve γ_n converges smoothly on $[0,a]$ to a geodesic γ.

LEMMA 3.17. *Suppose there is no energy concentration for the sequence ϕ_{α_n} in the region $D_\delta(x_n)\setminus D_{\lambda_n R}(x_n)$, i.e. (3.20) holds, if the limiting neck of ϕ_{α_n} is a geodesic of infinite length, then for any given $a > 0$ and any fixed θ, $\phi_{\alpha_n}(se^{\sqrt{-1}\theta})$ converges to γ in $C^1([0,a])$. Moreover, we have*

$$\left\| \frac{r(s)}{\sqrt{\alpha_n - 1}}\left|\frac{\partial s}{\partial r}\right| - \mu^{1-t_1}\sqrt{\frac{E(w^1)}{\pi}} \right\|_{C^0([0,a])} \to 0, \ \textit{as } n\to\infty.$$

PROOF. Let

$$s(\lambda_n^{t_n^a}) = a.$$

By Lemma 3.15, we have

$$a = \int_{\lambda_n^{t_1}}^{\lambda_n^{t_n^a}} \left|\frac{d\phi^*_{\alpha_n}(r)}{dr}\right| dr \ge C\int_{\lambda_n^{t_1}}^{\lambda_n^{t_n^a}} \frac{\sqrt{\alpha_n - 1}}{r} dr = C(t_n^a - t_1)\sqrt{\alpha_n - 1}\log\lambda_n. \tag{3.44}$$

On the other hand, Theorem 3.2 tells us

$$\sqrt{\alpha_n - 1}\log\lambda_n \to -\infty,$$

since the limiting neck geodesic is of infinite length. Hence, from above facts, we have

$$t_n^a - t_1 \to 0, \quad \text{as} \quad n\to +\infty. \tag{3.45}$$

We assume that $\phi_{\alpha_n}(se^{\sqrt{-1}\theta})$ does not converge to γ in $C^1[0,a]$. Then by Lemma 3.15, there exists $s_n \in [0,a]$, such that

$$\sup_\theta \left| \frac{\partial \phi_{\alpha_n}}{\partial s}(s_n e^{\sqrt{-1}\theta}) - \frac{d\phi^*_{\alpha_n}}{ds}(s_n) \right| > \epsilon > 0.$$

Let $s_n = s(\lambda_n^{t_n})$. It is easy to see that $t_n \in [t_n^a, t_1]$. Thus $t_n \to t_1$. By Lemma 3.15, after passing to a subsequence, we have

$$\lim_{n\to\infty} \frac{\lambda_n^{t_n}}{\sqrt{\alpha_n - 1}} \left| \frac{\partial \phi_{\alpha_n}}{\partial r}(\lambda_n^{t_n} e^{\sqrt{-1}\theta}) - \frac{d\phi^*_{\alpha_n}}{dr}(\lambda_n^{t_n}) \right| \to 0.$$

Therefore, noting

$$\left| \frac{\partial \phi_{\alpha_n}(s_n e^{\sqrt{-1}\theta})}{\partial s} - \frac{d\phi^*_{\alpha_n}(s_n)}{ds} \right| = \left| \frac{dr}{ds} \right| \cdot \left| \frac{\partial \phi_{\alpha_n}}{\partial r} - \frac{d\phi^*_{\alpha_n}(r)}{dr} \right|_{r=\lambda_n^{t_n}}$$

and

$$\left| \frac{dr}{ds} \right|_{r=\lambda_n^{t_n}} \leq \frac{C\lambda_n^{t_n}}{\sqrt{\alpha_n - 1}},$$

we have

$$\left| \frac{\partial \phi_{\alpha_n}(s_n e^{\sqrt{-1}\theta})}{\partial s} - \frac{d\phi^*_{\alpha_n}(s_n)}{ds} \right| \leq \frac{C\lambda_n^{t_n}}{\sqrt{\alpha_n - 1}} \left| \frac{\partial \phi_{\alpha_n}}{\partial r} - \frac{d\phi^*_{\alpha_n}(r)}{dr} \right|_{r=\lambda_n^{t_n}} \to 0,$$

which is a contradiction. Thus, we proved that $\phi_{\alpha_n}(se^{\sqrt{-1}\theta})$ converges to γ in $C^1([0,a])$

By similar arguments, we can prove

$$\left\| \frac{r(s)}{\sqrt{\alpha_n - 1}} \left| \frac{\partial s}{\partial r} \right| - \mu^{1-t_1} \sqrt{\frac{E(w^1)}{\pi}} \right\|_{C^0([0,a])} \to 0, \ as\ n \to \infty.$$

□

LEMMA 3.18. *Under the assumption of Theorem 3.2, if the limiting neck of ϕ_{α_n} is an unstable geodesic which is parameterized on $[0, a]$ by arc length, then, for $\alpha_n - 1 > 0$ sufficiently small, there exists a section V_n of $\phi^*_{\alpha_n}(TN)$, which is supported in $D_{\lambda_n^{t_n^a}}(x_n) \setminus D_{\lambda_n^{t_1}}(x_n)$, such that*

$$\delta^2 E_{\alpha_n}(\phi_{\alpha_n})(V_n, V_n) < 0.$$

PROOF. Since the limiting neck of $\{\phi_{\alpha_n}\}$, denoted by $\gamma : [0, a] \to N$, is not a stable geodesic, there exists a vector field V_0 on γ with $V_0|_{\gamma(0)} = 0$ and $V_0|_{\gamma(a)} = 0$ such that

$$I_\gamma(V_0, V_0) < 0.$$

Let P be the projection from $T\mathbb{R}^K$ to TN. We define

$$V_n\left(r(s)e^{\sqrt{-1}\theta} + x_n\right) = P_{\phi_{\alpha_n}(se^{\sqrt{-1}\theta})}(V_0(s)),$$

where s is the arc-length parametrization of $\phi^*_{\alpha_n}(r)$ with $s(\lambda_\alpha^{t_1}) = 0$. Then, $V_n(x)$ is a smooth section of $\phi^*_{\alpha_n}(TN)$ which is supported in $D_{\lambda_n^{t_n^a}}(x_n) \setminus D_{\lambda_n^{t_1}}(x_n)$. By

Lemma 3.17, for any fixed θ, $V_n(\phi_{\alpha_n}(se^{\sqrt{-1}\theta}))$ converges to $V_0(\gamma(s))$ in $C^1[0,b]$. Then by (2.57)

$$\begin{aligned}\delta^2 E_{\alpha_n}(V_n,V_n) = 2\alpha_n & \int_{D_{\lambda_n^{t_n^a}}\setminus D_{\lambda_n^{t_1}}} (1+|\nabla_{\tilde g_n}\phi_{\alpha_n}|^2)^{(\alpha_n-1)} \\ & \times (\langle \nabla V_n, \nabla V_n\rangle - R(V_n,\nabla\phi_{\alpha_n},\nabla\phi_{\alpha_n},V_n))\,dx \\ & + 4\alpha_n(\alpha_n-1)\int_{D_{\lambda_n^{t_n^a}}\setminus D_{\lambda_n^{t_1}}} (1+|\nabla_{\tilde g_n}\phi_{\alpha_n}|^2)^{\alpha_n-2}\langle\nabla\phi_{\alpha_n},\nabla V_n\rangle^2 dx \\ & := \mathbf{I}+\mathbf{II}. \end{aligned}\tag{3.46}$$

Next, we will show that

$$\lim_{n\to+\infty}\frac{1}{\sqrt{\alpha_n-1}}\delta^2 E_{\alpha_n}(V_n,V_n) = 4\pi\mu l_\infty\sqrt{\frac{E(v)}{\pi}}I_\gamma(V_0,V_0).\tag{3.47}$$

Firstly, we calculate $\mathbf{I}$:

$$\begin{aligned}\frac{\mathbf{I}}{\sqrt{\alpha_n-1}} & \\ = \frac{2\alpha_n}{\sqrt{\alpha_n-1}} & \int_{D_{\lambda_n^{t_n^a}}\setminus D_{\lambda_n^{t_1}}(x_n)} (1+|\nabla_{\tilde g_n}\phi_{\alpha_n}|^2)^{(\alpha_n-1)} \\ \times (\langle\nabla V_n, & \nabla V_n\rangle - R(V_n,\nabla\phi_{\alpha_n},\nabla\phi_{\alpha_n},V_n))\,dx \\ = 2\alpha_n\int_0^{2\pi} & \int_0^a (1+|\nabla_{\tilde g_n}\phi_{\alpha_n}|^2)^{(\alpha_n-1)} \\ \cdot\Big(\langle\nabla_{\frac{\partial\phi_{\alpha_n}}{\partial s}}V_n, & \nabla_{\frac{\partial\phi_{\alpha_n}}{\partial s}}V_n\rangle - R(V_n,\frac{\partial\phi_{\alpha_n}}{\partial s},\frac{\partial\phi_{\alpha_n}}{\partial s},V_n)\Big)\frac{|\frac{\partial s}{\partial r}|}{\sqrt{\alpha_n-1}}rdsd\theta \\ + \frac{2\alpha_n}{\sqrt{\alpha_n-1}} & \int_0^{2\pi}\int_{\lambda_n^{t_1}}^{\lambda_n^{t_n^a}} (1+|\nabla_{\tilde g_n}\phi_{\alpha_n}|^2)^{(\alpha_n-1)} \\ \cdot\Big(\langle\nabla_{r^{-1}\frac{\partial\phi_{\alpha_n}}{\partial\theta}}V_n, & \nabla_{r^{-1}\frac{\partial\phi_{\alpha_n}}{\partial\theta}}V_n\rangle - R(V_n,r^{-1}\frac{\partial\phi_{\alpha_n}}{\partial\theta},r^{-1}\frac{\partial\phi_{\alpha_n}}{\partial\theta},V_n)\Big)rdrd\theta \\ := \mathbf{I}_1+\mathbf{I}_2. & \end{aligned}$$

We claim that

$$\left(\left|\frac{\lambda_n^{t_n(s)}}{\sqrt{\alpha_n-1}}\nabla_{\tilde g_n}\phi_{\alpha_n}\right|^2\frac{\alpha_n-1}{\lambda_n^{2t_n(s)}}\right)^{(\alpha_n-1)} \longrightarrow l_\infty\mu^{t_1},\tag{3.48}$$

where $t_n^a\le t_n(s)\le t_1$. In fact, by Lemma 3.15 , we have

$$\lim_{n\to\infty}\left(\left|\frac{\lambda_n^{t_n(s)}}{\sqrt{\alpha_n-1}}\nabla_{\tilde g_n}\phi_{\alpha_n}\right|^2\frac{\alpha_n-1}{\lambda_n^{2t_n(s)}}\right)^{(\alpha_n-1)} = \lim_{n\to\infty}\left(\frac{1}{\tilde l_n^2\lambda_n^{2t_n(s)}}\right)^{(\alpha_n-1)}.\tag{3.49}$$

Since $\nu=\infty$, by Lemma 3.15, we have

$$a=\int_{\lambda_n^{t_2}}^{\lambda_n^{t_n^a}}|\phi^*_{\alpha_n}|dr \ge C\int_{\lambda_n^{t_2}}^{\lambda_n^{t_n^a}}\frac{\sqrt{\alpha_n-1}}{r}dr = -C(t_n^a-t_2)\sqrt{\alpha_n-1}\log\lambda_n.\tag{3.50}$$

Thus, we obtain that

$$t_n^a\to t_1.$$

This implies that (3.48) holds.

It follows from the fact that $\mu \geq 1$, $l_\infty \geq 1$ and (3.48) that

$$(1+|\nabla_{\tilde{g}_n}\phi_{\alpha_n}|^2)^{(\alpha_n-1)} = \left(1+\left|\frac{\lambda_n^{t_n(s)}}{\sqrt{\alpha_n-1}}\nabla_{\tilde{g}_n}\phi_{\alpha_n}\right|^2 \frac{\alpha_n-1}{\lambda_n^{2t_n(s)}}\right)^{(\alpha_n-1)} \longrightarrow \mu^{t_1}l_\infty.$$

Hence, we infer from the above fact and Lemma 3.17 that

$$\lim_{\alpha\searrow 1}\frac{\mathbf{I}_1}{\sqrt{\alpha-1}} = 4\pi\mu l_\infty\sqrt{\frac{E(w^1)}{\pi}}I_\gamma(V_0,V_0).$$

For the term $\mathbf{I}_2$, by definition, we have

$$\nabla_{\frac{\partial\phi_{\alpha_n}}{\partial\theta}}V_n = P_{\phi_{\alpha_n}(se^{\sqrt{-1}\theta})}\left(\frac{\partial V_n}{\partial\theta}\right) = P_{\phi_{\alpha_n}(se^{\sqrt{-1}\theta})}\left(\frac{\partial}{\partial\theta}(P_{\phi_{\alpha_n}(se^{\sqrt{-1}\theta})})(V_0)\right),$$

where $\frac{\partial V_n}{\partial\theta}$ is the derivative in $\mathbb{R}^K$. This leads to

$$|\nabla_{\frac{\partial\phi_{\alpha_n}}{\partial\theta}}V_n| \leq C(N)\left|\frac{\partial\phi_{\alpha_n}}{\partial\theta}\right|.$$

Hence, we have

$$\begin{aligned}\mathbf{I}_2 &= n2\alpha_n\int_0^{2\pi}\int_{\lambda_n^{t_1}}^{\lambda_n^{t_n^a}}\frac{(1+|\nabla_{\tilde{g}_n}\phi_{\alpha_n}|^2)^{(\alpha_n-1)}}{\sqrt{\alpha_n-1}}\\ &\quad\cdot\left(\langle\nabla_{r^{-1}\frac{\partial\phi_{\alpha_n}}{\partial\theta}}V_n,\nabla_{r^{-1}\frac{\partial\phi_{\alpha_n}}{\partial\theta}}V_n\rangle - R(V_n,r^{-1}\frac{\partial\phi_{\alpha_n}}{\partial\theta},r^{-1}\frac{\partial\phi_{\alpha_n}}{\partial\theta},V_n)\right)rdrd\theta\\ &\leq \frac{C}{\sqrt{\alpha_n-1}}\int_0^{2\pi}\int_{\lambda_n^{t_1}}^{\lambda_n^{t_n^a}}|r^{-1}\frac{\partial\phi_{\alpha_n}}{\partial\theta}|^2rdrd\theta.\end{aligned}$$

For a given $K>1$, set

$$m_n = \left[\frac{\log\lambda_n^{t_n^a-t_1}}{\log K}\right]+1.$$

It is easy to see that

$$D_{\lambda_n^{t_n^a}}\setminus D_{\lambda_n^{t_1}} \subset \cup_{i=1}^{m_n}(D_{K^i\lambda_n^{t_1}}\setminus D_{K^{i-1}\lambda_n^{t_1}}).$$

By (3.50), there holds

$$\sqrt{\alpha-1}\,m_n \leq C(K).$$

Then according to Lemma 3.13, we obtain

$$\begin{aligned}\mathbf{I}_2 &\leq \frac{C}{\sqrt{\alpha_n-1}}\int_0^{2\pi}\int_{\lambda_n^{t_1}}^{\lambda_n^{t_n^a}}|r^{-1}\frac{\partial\phi_{\alpha_n}}{\partial\theta}|^2rdrd\theta\\ &\leq \frac{C\sqrt{\alpha_n-1}m_n}{\alpha_n-1}\frac{1}{m_n}\int_{\cup_{i=1}^{m_n}\left(D_{K^i\lambda_n^{t_1}}\setminus D_{K^{i-1}\lambda_n^{t_1}}\right)}|r^{-1}\frac{\partial\phi_{\alpha_n}}{\partial\theta}|^2dx\\ &\leq \frac{C(K)}{\alpha_n-1}\frac{1}{m_n}\int_{\cup_{i=1}^{m_n}\left(D_{K^i\lambda_n^{t_1}}\setminus D_{K^{i-1}\lambda_n^{t_1}}\right)}|r^{-1}\frac{\partial\phi_{\alpha_n}}{\partial\theta}|^2dx \to 0. \qquad (3.51)\end{aligned}$$

Now, we turn to the term **II**. It is easy to check that

$$|\langle\nabla_{\tilde{g}_n}\phi_{\alpha_n},\nabla_{\tilde{g}_n}V_n\rangle| \leq C|\nabla_{\tilde{g}_n}\phi_{\alpha_n}|^2.$$

So, there exists a constant $C > 0$ such that

$$(1+|\nabla_{\tilde{g}_n}\phi_{\alpha_n}|^2)^{\alpha_n-2}\langle\nabla_{\tilde{g}_n}\phi_{\alpha_n},\nabla_{\tilde{g}_n}V_n\rangle^2 \le C(1+|\nabla_{\tilde{g}_n}\phi_{\alpha_n}|^2)^{\alpha_n}.$$

This leads to

$$\frac{\mathbf{II}}{\sqrt{\alpha_n-1}} \le C\sqrt{\alpha_n-1}\int_M (1+|\nabla_{\tilde{g}_n}\phi_{\alpha_n}|^2)^{\alpha_n} dvol_g \to 0.$$

Then we obtained the desired estimate and the conclusion of the lemma follows immediately. □

PROPOSITION 3.19. *Let (N, g_N) be a compact Riemannian manifold such that one of the following two assumptions holds:*

(i) *the Ricci curvature of the target manifold (N, g_N) has a positive lower bound, i.e. there exists a positive constant $\lambda_N > 0$ such that $Ric_N \ge \lambda_N > 0$;*
(ii) *the target manifold (N, g_N) has a finite fundamental group.*

Let $\phi_{\alpha_n} : (D_1(0), \tilde{g}_n) \to (N, h)$ be a sequence of α-harmonic maps with uniformly bounded α-energy. Suppose there is only one energy concentration point 0 in $D_1(0)$ for the sequence ϕ_{α_n} and there is only one bubble occurring at this point, if $\nu = \infty$, then the Morse index of ϕ_{α_n} tends to infinity.

PROOF. Similarly to the proof of Proposition 2.17, we divide the proof into two steps corresponding to the two assumptions.

STEP 1. the Ricci curvature of the target manifold (N, g_N) has a positive lower bound.

Since $\mathrm{Ric}_N \ge \lambda_N > 0$, by Myers' theorem, we know that, if

$$a \ge \frac{\pi}{\sqrt{\lambda_N(n-1)^{-1}}} + 2\epsilon,$$

then there exists a tangent vector field $V_0(s)$ on N, which is smooth on γ, and is vanishing on $\gamma|_{[0,\epsilon]}$ and $\gamma|_{[a-\epsilon,a]}$, such that the second variation of the length of γ satisfies

$$I_\gamma(V_0,V_0) = \int_0^a (\langle\nabla_{\dot\gamma}V_0,\nabla_{\dot\gamma}V_0\rangle - R(V_0,\dot\gamma,\dot\gamma,V_0))ds < -\delta < 0. \tag{3.52}$$

Since $\nu = \infty$, Theorem 3.2 tells us that the limiting neck of $\{\phi_{\alpha_n}\}$ is a geodesic of infinite length. Therefore, for any given t_1, by Lemma 3.15, we know that there exists a suitable positive constant $\epsilon_1 > 0$ such that, as $\alpha_n - 1 > 0$ is small enough, the arc length a of $\gamma_n(s)$ on the region $D_{\lambda_n^{t_1-\epsilon}}(x_n) \setminus D_{\lambda_n^{t_1}}(x_n)$ satisfies

$$a > \frac{\pi}{\sqrt{\lambda_N(n-1)^{-1}}}.$$

According to Lemma 3.18, there exists a section V_n^1 of $\phi_{\alpha_n}^*(TN)$, which is 0 outside the region $D_{\lambda_n^{t_1-\epsilon}}(x_n) \setminus D_{\lambda_n^{t_1}}(x_n)$, satisfying

$$\delta^2 E_{\alpha_n}(\phi_{\alpha_n})(V_n^1, V_n^1) < 0.$$

By the same method, for $t_2 = t_1 - 2\epsilon_1$, we can also pick $\epsilon_2 > 0$ and construct a section V_n^2, which is 0 outside $D_{\lambda_n^{t_2-\epsilon_2}}(x_n) \setminus D_{\lambda_n^{t_2}}(x_n)$, such that

$$\delta^2 E_{\alpha_n}(\phi_{\alpha_n})(V_n^2, V_n^2) < 0.$$

Since the limiting neck is a geodesic of infinite length, then, when $\alpha_n - 1 > 0$ is sufficiently small, there exists a series of sections $\{V_n^3, V_n^4, \cdots, V_n^k\}$ satisfying that for any $1 \le i \le k$ there holds

$$\delta^2 E_{\alpha_n}(\phi_{\alpha_n})(V_n^i, V_n^i) < 0.$$

Obviously, V_n^1, V_n^2, $\cdots$, V_n^k are linearly independent. This means that

$$\text{Index } (E_{\alpha_n}; \phi_{\alpha_n}) \ge k.$$

Therefore, we get

$$\text{Index } (E_{\alpha_n}; \phi_{\alpha_n}) \to +\infty, \quad \text{as} \quad \alpha_n \searrow 1.$$

STEP 2. the target manifold (N, g_N) has a finite fundamental group.

We prove this proposition by contradiction. If it is false, then there exists an uniform integer $\overline{m}$ such that Index $(\phi_{\alpha_n}; E_{\alpha_n}) \le \overline{m} < \infty$. Let $s = s(r)$ be the arc-length parameter as before with $s(\lambda_n^{t_1}) = 0$, $s(t_n^a) = a$ and $s(s_n^a) = -a$, where $0 < t_n^a < t^1 < s_n^a < 1$. By Theorem 3.2, $\gamma_n(s)$ converges to γ in $C^1([-a, a])$.

Next, we will show that the Morse index of γ on any subinterval $[-a, a]$ is uniformly bounded.

In fact, let $V_1(s), \ldots, V_m(s)$ be the m linearly independent tangent vector fields on N, which are smooth on γ, and are vanishing on $\gamma|_{[-a,-a+\epsilon]}$ and $\gamma|_{[a-\epsilon,a]}$ for any big $a > 0$ and small $\epsilon > 0$, such that for any $q \in \{1, \ldots, m\}$, the second variation of the length of γ satisfies

$$I_\gamma(V_q, V_q) = \int_{-b}^{b} (\langle \nabla_{\dot\gamma} V_q, \nabla_{\dot\gamma} V_q \rangle - R(V_q, \dot\gamma, \dot\gamma, V_q))ds < -\delta < 0. \tag{3.53}$$

Let P be projection from $T\mathbb{R}^K$ to TN. We define

$$V_n^q(s, \theta) = P_{\phi_{\alpha_n}(s,\theta)}(V_q(s)).$$

Similarly to the proof of Lemma 3.18, we have

$$\delta^2 E_{\alpha_n}(\phi_{\alpha_n})(V_n^q, V_n^q) < 0,$$

when $\alpha_n - 1 > 0$ is small enough.

Thus,

$$m \le \text{ Index } (\phi_{\alpha_n}; E_{\alpha_n}) \le \overline{m} < \infty,$$

which implies that the Morse index of the geodesic γ is uniformly bounded by $\overline{m}$, since a is arbitrary.

Since the target manifold N has a finite fundamental group, by Gromov estimates (see Corollary 3.3.5 in [**20**]), the length of the geodesic is controlled by the bound of the Morse index, i.e.,

$$\text{length}(\gamma) \le C\,(\text{Index}(\gamma) + 1) \le C(\overline{m} + 1) < \infty,$$

which is a contradiction to $\nu = \infty$.

Thus, we complete the proof of the proposition. □

At the end of this section, we prove Theorem 3.3.

Proof of Theorem 3.3. By Proposition 3.19, we know that the limiting necks consist of some finite length geodesics. Then Theorem 3.2 implies that $\mu = 1$. Thus it is easy to see that the conclusions of Theorem 3.3 hold. □

CHAPTER 4

Proof of main theorems

In this section, we will prove our main results: Theorem 1.4 - Theorem 1.6.

Proof of Theorem 1.4. We first prove that for each $j = 1, \ldots, p$, there holds

$$\begin{aligned} &\lim_{n\to\infty} \int_{[-T_n^j, T_n^j]\times S^1} \left((1+|\nabla_{g_n^j}\phi_{\alpha_n}|^2)^{\alpha_n} - 1\right) dvol_{g_n^j} \\ &= \sum_{k=1}^{K_j} l_{jk}^\infty E(\psi_{jk}) + \sum_{k=1}^{K_j} l_{jk}^\infty \sum_{l=1}^{L_{jk}} (\mu_{jk}^l)^2 E(w_{jk}^l) + \lim_{n\to\infty} \kappa_n^j \frac{4\pi^2}{l_n^j}. \end{aligned} \tag{4.1}$$

In fact, by the decomposition of $Q_n^j = [-T_n^j, T_n^j] \times S^1$ given in Section 1, we have

$$\begin{aligned} &\lim_{n\to\infty} \int_{[-T_n^j, T_n^j]\times S^1} \left((1+|\nabla_{g_n^j}\phi_{\alpha_n}|^2)^{\alpha_n} - 1\right) dvol_{g_n^j} \\ &= \lim_{n\to\infty} \sum_{i=0}^{K_j} \int_{I_n^{j,i}} \left((1+|\nabla_{g_n^j}\phi_{\alpha_n}|^2)^{\alpha_n} - 1\right) dvol_{g_n^j} \\ &\quad + \lim_{n\to\infty} \sum_{k=1}^{K_j} \int_{J_n^{j,k}} \left((1+|\nabla_{g_n^j}\phi_{\alpha_n}|^2)^{\alpha_n} - 1\right) dvol_{g_n^j} \\ &= \mathbf{I} + \mathbf{II}. \end{aligned}$$

For **I**, by (1.6) and Theorem 2.1, we have

$$\begin{aligned} \mathbf{I} = \lim_{n\to\infty} \sum_{i=0}^{K_j} 2\kappa_n^j |I_n^{j,i}| &= \lim_{n\to\infty} 2\kappa_n^j \left(2T_n^j - \sum_{i=1}^{K_j} (b_n^{j,i} - a_n^{j,i})\right) \\ &= \lim_{n\to\infty} 2\frac{\kappa_n^j}{T_n^j} \left(2 - \sum_{i=1}^{K_j} \frac{b_n^{j,i} - a_n^{j,i}}{T_n^j}\right) \\ &= \lim_{n\to\infty} 4\frac{\kappa_n^j}{T_n^j} = \kappa_n^j \frac{4\pi^2}{l_n^j}. \end{aligned} \tag{4.2}$$

For **II**, by (2.25), we first have

(4.3)

$$\lim_{n\to\infty} \sum_{k=1}^{K_j} \int_{J_n^{j,k}} dvol_{g_n^j} \le \lim_{R\to\infty} \lim_{n\to\infty} \int_{[-T_n^j+R, T_n^j-R]\times S^1} (\rho_n^j)^2 dt d\theta \le \lim_{R\to\infty} \frac{C}{R} = 0.$$

Denoting

$$S_{jk}^\delta := \cup_{x\in S_{jk}} D_\delta(x)$$

and

$$\overline{g}_n^j(t,\theta) = g_n^j(t + \frac{b_n^{j,k} + a_n^{j,k}}{2}, \theta), \quad \overline{\phi}_{\alpha_n}(t,\theta) = \phi_{\alpha_n}(t + \frac{b_n^{j,k} + a_n^{j,k}}{2}, \theta),$$

then we have

$$\begin{aligned}
\mathbf{II} &= \lim_{n\to\infty} \sum_{k=1}^{K_j} \int_{J_n^{j,k}} \Big((1 + |\nabla_{g_n^j} \phi_{\alpha_n}|^2)^{\alpha_n} - 1\Big)\, dvol_{g_n^j} \\
&= \lim_{n\to\infty} \sum_{k=1}^{K_j} \int_{J_n^{j,k}} \Big((1 + |\nabla_{g_n^j} \phi_{\alpha_n}|^2)^{\alpha_n} - 1\Big)\, dvol_{g_n^j} \\
&= \lim_{n\to\infty} \sum_{k=1}^{K_j} \int_{[-\frac{b_n^{j,k} - a_n^{j,k}}{2}, \frac{b_n^{j,k} - a_n^{j,k}}{2}]\times S^1} \Big((1 + |\nabla_{\overline{g}_n^j} \overline{\phi}_{\alpha_n}|^2)^{\alpha_n} - 1\Big)\, dvol_{\overline{g}_n^j} \\
&= \lim_{\delta\to 0}\lim_{n\to\infty} \sum_{k=1}^{K_j} \int_{[-\frac{b_n^{j,k} - a_n^{j,k}}{2}, \frac{b_n^{j,k} - a_n^{j,k}}{2}]\times S^1 \setminus S_{jk}^\delta} \Big((1 + |\nabla_{\overline{g}_n^j} \overline{\phi}_{\alpha_n}|^2)^{\alpha_n} - 1\Big)\, dvol_{\overline{g}_n^j} \\
&\quad + \lim_{\delta\to 0}\lim_{n\to\infty} \sum_{k=1}^{K_j} \int_{S_{jk}^\delta} \Big((1 + |\nabla_{\overline{g}_n^j} \overline{\phi}_{\alpha_n}|^2)^{\alpha_n} - 1\Big)\, dvol_{\overline{g}_n^j} \\
&= \lim_{R\to\infty}\lim_{n\to\infty} \sum_{k=1}^{K_j} \int_{[-\frac{b_n^{j,k} - a_n^{j,k}}{2}, -R]\times S^1} \Big((1 + |\nabla_{\overline{g}_n^j} \overline{\phi}_{\alpha_n}|^2)^{\alpha_n} - 1\Big)\, dvol_{\overline{g}_n^j} \\
&\quad + \lim_{R\to\infty}\lim_{n\to\infty} \sum_{k=1}^{K_j} \int_{[R, \frac{b_n^{j,k} - a_n^{j,k}}{2}]\times S^1} \Big((1 + |\nabla_{\overline{g}_n^j} \overline{\phi}_{\alpha_n}|^2)^{\alpha_n} - 1\Big)\, dvol_{\overline{g}_n^j} \\
&\quad + \lim_{R\to\infty}\lim_{\delta\to 0}\lim_{n\to\infty} \sum_{k=1}^{K_j} \int_{[-R,R]\times S^1 \setminus S_{jk}^\delta} \Big((1 + |\nabla_{\overline{g}_n^j} \overline{\phi}_{\alpha_n}|^2)^{\alpha_n} - 1\Big)\, dvol_{\overline{g}_n^j} \\
&\quad + \lim_{\delta\to 0}\lim_{n\to\infty} \sum_{k=1}^{K_j} \int_{S_{jk}^\delta} \Big((1 + |\nabla_{\overline{g}_n^j} \overline{\phi}_{\alpha_n}|^2)^{\alpha_n} - 1\Big)\, dvol_{\overline{g}_n^j} := \mathbf{II}_1 + \mathbf{II}_2 + \mathbf{II}_3 + \mathbf{II}_4.
\end{aligned}$$

Noting that

$$\begin{aligned}
&(1 + |\nabla_{\overline{g}_n^j} \overline{\phi}_{\alpha_n}|^2)^{\alpha_n - 1} \le C(l_n^j)^{2-2\alpha_n} \le C(\delta), \\
&\forall\, x \in \left[-\frac{b_n^{j,k} - a_n^{j,k}}{2}, \frac{b_n^{j,k} + a_n^{j,k}}{2}\right] \times S^1 \setminus S_{jk}^\delta,
\end{aligned}$$

by (1.5), (1.7) and (1.9), we get

$$\begin{aligned}
\mathbf{II}_1 \leq & \lim_{R\to\infty}\lim_{n\to\infty}\sum_{k=1}^{K_j}\int_{[-\frac{b_n^{j,k}-a_n^{j,k}}{2},-R]\times S^1}(1+|\nabla_{\overline{g}_n^j}\overline{\phi}_{\alpha_n}|^2)^{\alpha_n}dvol_{\overline{g}_n^j}\\
&+\lim_{R\to\infty}\lim_{n\to\infty}\sum_{k=1}^{K_j}\int_{[-\frac{b_n^{j,k}-a_n^{j,k}}{2},-R]\times S^1}dvol_{\overline{g}_n^j}\\
\leq & \lim_{R\to\infty}\lim_{n\to\infty}C\sum_{k=1}^{K_j}\int_{[-\frac{b_n^{j,k}-a_n^{j,k}}{2},-R]\times S^1}|\nabla_{\overline{g}_n^j}\overline{\phi}_{\alpha_n}|^2dvol_{\overline{g}_n^j}\\
&+\lim_{R\to\infty}\lim_{n\to\infty}\sum_{k=1}^{K_j}\int_{[-\frac{b_n^{j,k}-a_n^{j,k}}{2},-R]\times S^1}dvol_{\overline{g}_n^j}\\
\leq & \lim_{R\to\infty}\lim_{n\to\infty}C\sum_{k=1}^{K_j}\int_{[-\frac{b_n^{j,k}-a_n^{j,k}}{2},-R]\times S^1}|\nabla\overline{\phi}_{\alpha_n}|^2dtd\theta\\
&+\lim_{n\to\infty}\int_{[a_n^{j,1},b_n^{j,K_j}]\times S^1}dvol_{g_n^j}\\
\leq & \lim_{R\to\infty}\lim_{n\to\infty}C\sum_{k=1}^{K_j}\int_{[-\frac{b_n^{j,k}-a_n^{j,k}}{2},-R]\times S^1}|\nabla\overline{\phi}_{\alpha_n}|^2dtd\theta\\
&+\lim_{R\to\infty}\lim_{n\to\infty}\int_{[-T_n^j+R,T_n^j-R]\times S^1}dvol_{g_n^j}=0.
\end{aligned}$$

This implies $\mathbf{II}_1 = 0$. Similarly, we have $\mathbf{II}_2 = 0$.

For $\mathbf{II}_3$, noting that

$$\lim_{n\to\infty}\Big((1+|\nabla_{\overline{g}_n^j}\overline{\phi}_{\alpha_n}|^2)^{\alpha_n}-1\Big)=\lim_{n\to\infty}|\nabla_{\overline{g}_n^j}\overline{\phi}_{\alpha_n}|^{2\alpha_n},\quad \forall\, x\in[-R,R]\times S^1\setminus S_{jk}^\delta,$$

we have

$$\begin{aligned}
\mathbf{II}_3 = & \lim_{R\to\infty}\lim_{\delta\to0}\lim_{n\to\infty}\sum_{k=1}^{K_j}\int_{[-R,R]\times S^1\setminus S_{jk}^\delta}|\nabla_{\overline{g}_n^j}\overline{\phi}_{\alpha_n}|^{2\alpha_n}dvol_{\overline{g}_n^j}\\
= & \lim_{R\to\infty}\lim_{\delta\to0}\lim_{n\to\infty}\sum_{k=1}^{K_j}\int_{[-R,R]\times S^1\setminus S_{jk}^\delta}|\nabla\overline{\phi}_{\alpha_n}|^{2\alpha_n}\\
&\times\left(\frac{l_n^j}{2\pi\cos(\frac{l_n^j}{2\pi}(\frac{a_n^{j,k}+b_n^{j,k}}{2}+t))}\right)^{2-2\alpha_n}dtd\theta\\
= & \lim_{R\to\infty}\lim_{\delta\to0}\lim_{n\to\infty}\sum_{k=1}^{K_j}\int_{[-R,R]\times S^1\setminus S_{jk}^\delta}|\nabla\overline{\phi}_{\alpha_n}|^{2\alpha_n}\\
&\times\left(\frac{l_n^j}{2\pi\cos(\frac{l_n^j}{2\pi}(\frac{a_n^{j,k}+b_n^{j,k}}{2}))}\right)^{2-2\alpha_n}dtd\theta\\
= & \lim_{R\to\infty}\lim_{\delta\to0}\sum_{k=1}^{K_j}l_{jk}^\infty\int_{[-R,R]\times S^1\setminus S_{jk}^\delta}|\nabla\psi_{jk}|^2dtd\theta
\end{aligned}$$

$$= \sum_{k=1}^{K_j} l_{jk}^{\infty} E(\psi_{jk}),$$

where the third equality follows from the fact that

$$\lim_{n\to\infty} \left(\frac{l_n^j}{2\pi\cos(\frac{l_n^j}{2\pi}(\frac{a_n^{j,k}+b_n^{j,k}}{2}+t))} \right)^{2-2\alpha_n} = \left(\frac{l_n^j}{2\pi\cos(\frac{l_n^j}{2\pi}(\frac{a_n^{j,k}+b_n^{j,k}}{2}))} \right)^{2-2\alpha_n} \quad \forall\, t\in[-R,R]. \tag{4.4}$$

For $\mathbf{II}_4$, by (1.7) (4.4) and Theorem 3.1, we get

$$\begin{aligned}
\mathbf{II}_4 &= \lim_{\delta\to 0}\lim_{n\to\infty} \sum_{k=1}^{K_j} \int_{S_{jk}^{\delta}} \left((1+|\nabla_{\overline{g}_n^j}\overline{\phi}_{\alpha_n}|^2)^{\alpha_n} - 1 \right) dvol_{\overline{g}_n^j} \\
&= \lim_{\delta\to 0}\lim_{n\to\infty} \sum_{k=1}^{K_j} \int_{S_{jk}^{\delta}} (1+|\nabla_{\overline{g}_n^j}\overline{\phi}_{\alpha_n}|^2)^{\alpha_n} dvol_{\overline{g}_n^j} \\
&= \sum_{k=1}^{K_j} l_{jk}^{\infty} \sum_{l=1}^{L_{jk}} (\mu_{jk}^l)^2 E(w_{jk}^l)
\end{aligned}$$

and then (4.1) follows.

Secondly, near the blow-up points x_i, $i = 1, \ldots, I$ in the non-degenerating region, by applying similar argument as in [**14**], we have

$$\lim_{\delta\to 0}\lim_{n\to\infty} E_{\alpha_n}(\phi_{\alpha_n}, D_\delta(x_i)) = \sum_{l=1}^{k_i} \mu_{il}^2 E(\xi_{il}).$$

Combining this with (4.1), the conclusion of this theorem follows immediately. □

Now we prove Theorem 1.5.

Proof of Theorem 1.5. Firstly, it is easy to see that conclusion (1) follows from [**14**] and (2) follows from Theorem 3.2. We just need to prove the conclusion (3).

For each $j = 1, \ldots, p$, by Theorem 2.2, we know that ϕ_{α_n} on $I_n^{j,i}$, $i = 0, 1, \ldots, K_j$ are converging to the geodesics on N of length

$$\frac{1}{\sqrt{4\pi}}\sqrt{\Xi_n^j}\,(a_n^{j,i+1} - b_n^{j,i}).$$

Then the sum of the lengths of the geodesics connecting $\phi(M)$ and $\psi_{jk}(S^2)$, $k = 1, \ldots, K_j$ near the jth pair of punctures is

$$\begin{aligned}
&\lim_{n\to\infty} \sum_{i=0}^{K_j} \frac{1}{\sqrt{4\pi}} \sqrt{\Xi_n^j} (a_n^{j,i+1} - b_n^{j,i}) \\
&= \lim_{n\to\infty} \frac{1}{\sqrt{4\pi}} \sqrt{\Xi_n^j} \left(2T_n^j - \sum_{i=1}^{K_j} (b_n^{j,i} - a_n^{j,i}) \right) \\
&= \lim_{n\to\infty} \frac{1}{\sqrt{4\pi}} \sqrt{\Xi_n^j T_n^j} \left(2 - \sum_{i=1}^{K_j} \frac{b_n^i - a_n^i}{T_n^j} \right) \\
&= \lim_{n\to\infty} \frac{1}{\sqrt{\pi}} \sqrt{\Xi_n^j T_n^j} = \lim_{n\to\infty} \frac{1}{\sqrt{\pi}} \sqrt{\Xi_n^j \frac{\pi^2}{l_n^j}},
\end{aligned}$$

which implies the conclusions of the theorem immediately. □

Finally, we prove Theorem 1.6.

Proof of Theorem 1.6. Under the assumption of

$$\lim_{n\to\infty} \frac{(1-\alpha_n)\log l_n}{(l_n)^2} < \infty,$$

it is easy to see that

$$\lim_{n\to\infty} (l_n^j)^{2-2\alpha_n} = 1$$

which implies

$$l_{jk}^\infty = 1, \quad j = 1, \ldots, p, \; k = 1, \ldots, K_j.$$

Then the conclusion of Theorem 1.6 follows from Theorem 2.3 and Theorem 3.3. We finished the proof. □

Bibliography

[1] J. Chen, A. Fraser, and C. Pang, *Minimal immersions of compact bordered Riemann surfaces with free boundary*, Trans. Amer. Math. Soc. **367** (2015), no. 4, 2487–2507, DOI 10.1090/S0002-9947-2014-05990-4. MR3301871

[2] J. Chen and Y. Li, *Homotopy classes of harmonic maps of the stratified 2-spheres and applications to geometric flows*, Adv. Math. **263** (2014), 357–388, DOI 10.1016/j.aim.2014.07.001. MR3239142

[3] J. Chen and G. Tian, *Compactification of moduli space of harmonic mappings*, Comment. Math. Helv. **74** (1999), no. 2, 201–237, DOI 10.1007/s000140050086. MR1691947

[4] L. Chen, Y. Li, and Y. Wang, *The refined analysis on the convergence behavior of harmonic map sequence from cylinders*, J. Geom. Anal. **22** (2012), no. 4, 942–963, DOI 10.1007/s12220-011-9222-z. MR2965357

[5] W. Ding and G. Tian, *Energy identity for a class of approximate harmonic maps from surfaces*, Comm. Anal. Geom. **3** (1995), no. 3-4, 543–554, DOI 10.4310/CAG.1995.v3.n4.a1. MR1371209

[6] A. M. Fraser, *On the free boundary variational problem for minimal disks*, Comm. Pure Appl. Math. **53** (2000), no. 8, 931–971, DOI 10.1002/1097-0312(200008)53:8⟨931::AID-CPA1⟩3.3.CO;2-0. MR1755947

[7] A. M. Fraser, *Fundamental groups of manifolds with positive isotropic curvature*, Ann. of Math. (2) **158** (2003), no. 1, 345–354, DOI 10.4007/annals.2003.158.345. MR1999925

[8] A. Fraser and J. Wolfson, *The fundamental group of manifolds of positive isotropic curvature and surface groups*, Duke Math. J. **133** (2006), no. 2, 325–334, DOI 10.1215/S0012-7094-06-13325-2. MR2225695

[9] J. Jost, *Two-dimensional geometric variational problems*, Pure and Applied Mathematics (New York), John Wiley & Sons, Ltd., Chichester, 1991. A Wiley-Interscience Publication. MR1100926

[10] T. Lamm, *Energy identity for approximations of harmonic maps from surfaces*, Trans. Amer. Math. Soc. **362** (2010), no. 8, 4077–4097, DOI 10.1090/S0002-9947-10-04912-3. MR2608396

[11] J. Jost, L. Liu, and M. Zhu, *Blow-up analysis for approximate Dirac-harmonic maps in dimension 2 with applications to the Dirac-harmonic heat flow*, Calc. Var. Partial Differential Equations **56** (2017), no. 4, Paper No. 108, 26, DOI 10.1007/s00526-017-1202-0. MR3669777

[12] J. Li and X. Zhu, *Energy identity and necklessness for a sequence of Sacks-Uhlenbeck maps to a sphere*, Ann. Inst. H. Poincaré C Anal. Non Linéaire **36** (2019), no. 1, 103–118, DOI 10.1016/j.anihpc.2018.04.002. MR3906867

[13] Y. Li, L. Liu, and Y. Wang, *Blow-up behavior of harmonic maps with finite index*, Calc. Var. Partial Differential Equations **56** (2017), no. 5, Paper No. 146, 16, DOI 10.1007/s00526-017-1211-z. MR3708270

[14] Y. Li and Y. Wang, *A weak energy identity and the length of necks for a sequence of Sacks-Uhlenbeck α-harmonic maps*, Adv. Math. **225** (2010), no. 3, 1134–1184, DOI 10.1016/j.aim.2010.03.020. MR2673727

[15] Y. Li and Y. Wang, *A counterexample to the energy identity for sequences of α-harmonic maps*, Pacific J. Math. **274** (2015), no. 1, 107–123, DOI 10.2140/pjm.2015.274.107. MR3320872

[16] L. Liu, *No neck for Dirac-harmonic maps*, Calc. Var. Partial Differential Equations **52** (2015), no. 1-2, 1–15, DOI 10.1007/s00526-013-0702-9. MR3299172

[17] M. J. Micallef and J. D. Moore, *Minimal two-spheres and the topology of manifolds with positive curvature on totally isotropic two-planes*, Ann. of Math. (2) **127** (1988), no. 1, 199–227, DOI 10.2307/1971420. MR924677

[18] W. H. Meeks III and S. T. Yau, *Topology of three-dimensional manifolds and the embedding problems in minimal surface theory*, Ann. of Math. (2) **112** (1980), no. 3, 441–484, DOI 10.2307/1971088. MR595203

[19] J. D. Moore, *Energy growth in minimal surface bubbles*, unpublished manuscript, 2007.

[20] J. D. Moore, *Introduction to global analysis: Minimal surfaces in Riemannian manifolds*, Graduate Studies in Mathematics, vol. 187, American Mathematical Society, Providence, RI, 2017, DOI 10.1090/gsm/187. MR3729450

[21] T. H. Parker, *Bubble tree convergence for harmonic maps*, J. Differential Geom. **44** (1996), no. 3, 595–633. MR1431008

[22] J. Qing and G. Tian, *Bubbling of the heat flows for harmonic maps from surfaces*, Comm. Pure Appl. Math. **50** (1997), no. 4, 295–310, DOI 10.1002/(SICI)1097-0312(199704)50:4⟨295::AID-CPA1⟩3.0.CO;2-5. MR1438148

[23] J. Sacks and K. Uhlenbeck, *The existence of minimal immersions of 2-spheres*, Ann. of Math. (2) **113** (1981), no. 1, 1–24, DOI 10.2307/1971131. MR604040

[24] J. Sacks and K. Uhlenbeck, *Minimal immersions of closed Riemann surfaces*, Trans. Amer. Math. Soc. **271** (1982), no. 2, 639–652, DOI 10.2307/1998902. MR654854

[25] R. Schoen and S. T. Yau, *Existence of incompressible minimal surfaces and the topology of three-dimensional manifolds with nonnegative scalar curvature*, Ann. of Math. (2) **110** (1979), no. 1, 127–142, DOI 10.2307/1971247. MR541332

[26] Y. T. Siu and S. T. Yau, *Compact Kähler manifolds of positive bisectional curvature*, Invent. Math. **59** (1980), no. 2, 189–204, DOI 10.1007/BF01390043. MR577360

[27] M. Struwe, *On a free boundary problem for minimal surfaces*, Invent. Math. **75** (1984), no. 3, 547–560, DOI 10.1007/BF01388643. MR735340

[28] M. Zhu, *Harmonic maps from degenerating Riemann surfaces*, Math. Z. **264** (2010), no. 1, 63–85, DOI 10.1007/s00209-008-0452-0. MR2564932

Editorial Information

To be published in the *Memoirs*, a paper must be correct, new, nontrivial, and significant. Further, it must be well written and of interest to a substantial number of mathematicians. Piecemeal results, such as an inconclusive step toward an unproved major theorem or a minor variation on a known result, are in general not acceptable for publication.

Papers appearing in *Memoirs* are generally at least 80 and not more than 200 published pages in length. Papers less than 80 or more than 200 published pages require the approval of the Managing Editor of the Transactions/Memoirs Editorial Board. Published pages are the same size as those generated in the style files provided for $\mathcal{A}_{\mathcal{M}}\mathcal{S}$-LaTeX.

Information on the backlog for this journal can be found on the AMS website starting from `http://www.ams.org/memo`.

A Consent to Publish is required before we can begin processing your paper. After a paper is accepted for publication, the Providence office will send a Consent to Publish and Copyright Agreement to all authors of the paper. By submitting a paper to the *Memoirs*, authors certify that the results have not been submitted to nor are they under consideration for publication by another journal, conference proceedings, or similar publication.

Information for Authors

Memoirs is an author-prepared publication. Once formatted for print and on-line publication, articles will be published as is with the addition of AMS-prepared frontmatter and backmatter. Articles are not copyedited; however, confirmation copy will be sent to the authors.

Initial submission. The AMS uses Centralized Manuscript Processing for initial submissions. Authors should submit a PDF file using the Initial Manuscript Submission form found at `www.ams.org/submission/memo`, or send one copy of the manuscript to the following address: Centralized Manuscript Processing, MEMOIRS OF THE AMS, 201 Charles Street, Providence, RI 02904-2294 USA. If a paper copy is being forwarded to the AMS, indicate that it is for *Memoirs* and include the name of the corresponding author, contact information such as email address or mailing address, and the name of an appropriate Editor to review the paper (see the list of Editors below).

The paper must contain a *descriptive title* and an *abstract* that summarizes the article in language suitable for workers in the general field (algebra, analysis, etc.). The *descriptive title* should be short, but informative; useless or vague phrases such as "some remarks about" or "concerning" should be avoided. The *abstract* should be at least one complete sentence, and at most 300 words. Included with the footnotes to the paper should be the 2020 *Mathematics Subject Classification* representing the primary and secondary subjects of the article. The classifications are accessible from `www.ams.org/msc/`. The Mathematics Subject Classification footnote may be followed by a list of *key words and phrases* describing the subject matter of the article and taken from it. Journal abbreviations used in bibliographies are listed in the latest *Mathematical Reviews* annual index. The series abbreviations are also accessible from `www.ams.org/msnhtml/serials.pdf`. To help in preparing and verifying references, the AMS offers MR Lookup, a Reference Tool for Linking, at `www.ams.org/mrlookup/`.

Electronically prepared manuscripts. The AMS encourages electronically prepared manuscripts, with a strong preference for $\mathcal{A}_{\mathcal{M}}\mathcal{S}$-LaTeX. To this end, the Society has prepared $\mathcal{A}_{\mathcal{M}}\mathcal{S}$-LaTeX author packages for each AMS publication. Author packages include instructions for preparing electronic manuscripts, samples, and a style file that generates the particular design specifications of that publication series.

Authors may retrieve an author package for *Memoirs of the AMS* from `www.ams.org/journals/memo/memoauthorpac.html`. The *AMS Author Handbook* is available in PDF format from the author package link. The author package can also be obtained free of charge by sending email to `tech-support@ams.org` or from the Publication Division,

American Mathematical Society, 201 Charles St., Providence, RI 02904-2294, USA. When requesting an author package, please specify the publication in which your paper will appear. Please be sure to include your complete mailing address.

After acceptance. The source files for the final version of the electronic manuscript should be sent to the Providence office immediately after the paper has been accepted for publication. The author should also submit a PDF of the final version of the paper to the editor, who will forward a copy to the Providence office.

Accepted electronically prepared files can be submitted via the web at www.ams.org/submit-book-journal/, sent via FTP, or sent on CD to the Electronic Prepress Department, American Mathematical Society, 201 Charles Street, Providence, RI 02904-2294 USA. TEX source files and graphic files can be transferred over the Internet by FTP to the Internet node ftp.ams.org (130.44.1.100). When sending a manuscript electronically via CD, please be sure to include a message indicating that the paper is for the *Memoirs*.

Electronic graphics. Comprehensive instructions on preparing graphics are available at www.ams.org/authors/journals.html. A few of the major requirements are given here.

Submit files for graphics as EPS (Encapsulated PostScript) files. This includes graphics originated via a graphics application as well as scanned photographs or other computer-generated images. If this is not possible, TIFF files are acceptable as long as they can be opened in Adobe Photoshop or Illustrator.

Authors using graphics packages for the creation of electronic art should also avoid the use of any lines thinner than 0.5 points in width. Many graphics packages allow the user to specify a "hairline" for a very thin line. Hairlines often look acceptable when proofed on a typical laser printer. However, when produced on a high-resolution laser imagesetter, hairlines become nearly invisible and will be lost entirely in the final printing process.

Screens should be set to values between 15% and 85%. Screens which fall outside of this range are too light or too dark to print correctly. Variations of screens within a graphic should be no less than 10%.

Any graphics created in color will be rendered in grayscale for the printed version unless color printing is authorized by the Managing Editor and the Publisher. In general, color graphics will appear in color in the online version.

Inquiries. Any inquiries concerning a paper that has been accepted for publication should be sent to memo-query@ams.org or directly to the Electronic Prepress Department, American Mathematical Society, 201 Charles St., Providence, RI 02904-2294 USA.

Editors

This journal is designed particularly for long research papers, normally at least 80 pages in length, and groups of cognate papers in pure and applied mathematics. Papers intended for publication in the *Memoirs* should be addressed to one of the following editors. The AMS uses Centralized Manuscript Processing for initial submissions to AMS journals. Authors should follow instructions listed on the Initial Submission page found at www.ams.org/memo/memosubmit.html.

1520 **Guixiang Hong, Simeng Wang, and Xumin Wang,** Pointwise Convergence of Noncommutative Fourier Series, 2024

1519 **Katarzyna Mazowiecka, Michał Miśkiewicz, and Armin Schikorra,** On the Size of the Singular Set of Minimizing Harmonic Maps, 2024

1518 **L. Alexander Betts,** The Motivic Anabelian Geometry of Local Heights on Abelian Varieties, 2024

1517 **Changningphaabi Namoijam and Matthew A. Papanikolas,** Hyperderivatives of Periods and Quasi-Periods for Anderson t-Modules, 2024

1516 **Daniel Han-Kwan and David Michel,** On Hydrodynamic Limits of the Vlasov–Navier–Stokes System, 2024

1515 **Chenchen Mou and Jianfeng Zhang,** Wellposedness of Second Order Master Equations for Mean Field Games with Nonsmooth Data, 2024

1514 **Yohann Le Floch and Joseph Palmer,** Semitoric Families, 2024

1513 **Alcides Buss, Siegfried Echterhoff, and Rufus Willett,** Amenability and Weak Containment for Actions of Locally Compact Groups on C^*-Algebras, 2024

1512 **Alessandro Audrito and Susanna Terracini,** On the Nodal Set of Solutions to a Class of Nonlocal Parabolic Equations, 2024

1511 **Ermerson Araujo, Yuri Lima, and Mauricio Poletti,** Symbolic Dynamics for Nonuniformly Hyperbolic Maps with Singularities in High Dimension, 2024

1510 **C. Bleak, P. Cameron, Y. Maissel, A. Navas, and F. Olukoya,** The Further Chameleon Groups of Richard Thompson and Graham Higman: Automorphisms via Dynamics for the Higman-Thompson Groups $G_{n,r}$, 2024

1509 **Anthony Genevois,** Automorphisms of Graph Products of Groups and Acylindrical Hyperbolicity, 2024

1508 **Andrew J. Blumberg and Michael A. Mandell,** The Strong Künneth Theorem for Topological Periodic Cyclic Homology, 2024

1507 **Myoungjean Bae, Gui-Qiang G. Chen, and Mikhail Feldman,** Prandtl-Meyer Reflection Configurations, Transonic Shocks, and Free Boundary Problems, 2024

1506 **Radu Saghin, Pancho Valenzuela-Henríquez, and Carlos H. Vásquez,** Regularity with Respect to the Parameter of Lyapunov Exponents for Diffeomorphisms with Dominated Splitting, 2024

1505 **Deniz Bilman and Peter D. Miller,** Extreme Superposition: High-Order Fundamental Rogue Waves in the Far-Field Regime, 2024

1504 **Daniele Alessandrini, Olivier Guichard, Eugen Rogozinnikov, and Anna Wienhard,** Noncommutative Coordinates for Symplectic Representations, 2024

1503 **J. Adams, N. Arancibia Robert, and P. Mezo,** Equivalent Definitions of Arthur Packets for Real Classical Groups, 2024

1502 **R. M. Guralnick and R. Lawther,** Generic Stabilizers in Actions of Simple Algebraic Groups, 2024

1501 **Erasmo Caponio, Miguel Angel Javaloyes, and Miguel Sánchez,** Wind Finslerian Structures: From Zermelo's Navigation to the Causality of Spacetimes, 2024

1500 **Kenji Fukaya, Yong-Geun Oh, Hiroshi Ohta, and Kaoru Ono,** Exponential Decay Estimates and Smoothness of the Moduli Space of Pseudoholomorphic Curves, 2024

1499 **Stephen Theriault,** Homotopy Fibrations with a Section after Looping, 2024

1498 **Fabio Pusateri and Avraham Soffer,** Bilinear Estimates in the Presence of a Large Potential and a Critical NLS in 3D, 2024

1497 **Itay Glazer and Yotam I. Hendel,** On Singularity Properties of Word Maps and Applications to Probabilistic Waring Type Problems, 2024

For a complete list of titles in this series, visit the AMS Bookstore at **www.ams.org/bookstore/memoseries/**.